Macro-economic regulation

Macro-economic regulation

A control system incorporating the Kalman filter for the Netherlands

KESHAV P. VISHWAKARMA
Econometric Institute,
Erasmus University Rotterdam

Rotterdam University Press / 1974

Printed in the Netherlands

ISBN 90 237 2252 3

Preface

This study is the first to apply the elegant prediction and control algorithms developed by Kalman and Bucy (1961) to an economic problem. Why this lag of 12 years, and what made it end now?

The Kalman and Bucy paper was published in a journal widely read by control engineers but almost entirely ignored by economists. Since its appearance it has been quoted in general control texts (e.g. Aoki) but these in their turn have rarely been consulted by economists.

The few economists who took the trouble to look into control literature must, first of all, have been put off by an unfamiliar notation. In economic models, x stands for exogenous variables, and in control theory the x are endogenous. The u that economists keep for disturbances designates instruments in control. Language is different, too, and 'system variables', 'impulses' and 'controls' must be translated into endogenous variables, data, and instruments before the economist can begin to feel at home.

The strangest elements, however, are the 'signals' which economists should understand as statistical observations. It is only in the last few years that econometricians have re-introduced observation errors, thus allowing for discrepancies between the 'real' value of a variable and the observed value. Even so there remains a one-to-one relationship between the real and the observed values – both belong to the same (endogenous) variables. Dr. Vishwakarma shows that a much more general relationship between system variables and observations allows a better representation of the economic facts of life.

Econometricians entering control country are further put off by the almost complete lack of interest in estimation as understood in econometrics. Control engineers take their systems as 'given'. After all, most control systems are constructed in a known manner from components with known characteristics. It is true that economic policy analysts also behave as if they know how their systems operate, and they might therefore recognize some of their problems in control literature. But most policy analysts try to forget about uncertainty, while control theory is full of 'noise'.

This may explain why the present study has no known forerunners. How then did it come into being at this time?

There is no denying that Dr. Vishwakarma is a control expert invading economic territory. No economist will be deceived by his description of the economic system he experimented with. It is not the first time that economics has been visited by mathematicians, physicists, or engineers, and many of those visits have been very profitable for economics, particularly if the visitors had taken the trouble to learn the language of economics first. This study may well turn out to be another element in a series that starts with Cournot.

From a practical angle, Dr. Vishwakarma's methods promise answers to problems perpetually before those who must interpret tardy, incomplete, and often contradictory current statistics. Should one wait for more evidence to appear before drastic action is suggested, or would that mean that valuable time was lost? This study proposes precise answers to such questions. In practice, such answers will often be demonstrably wrong, or in any case unacceptable to those who have so far been doing not too badly by intuition alone. But a beginning has been made, and economists interested in decision problems will do well to read the signals this study sends forth.

J. Sandee

Contents

List of tables

List of figures

1. Introduction

1.1. SHORT-TERM ECONOMIC POLICY AND CONTROL THEORY

The objective of this study is to consider short-term economic policy from the viewpoint of modern control theory. The principal concern of short-term policy is with the regulation of economic activity so as to make it progress in a satisfactory manner, and it is largely of quantitative nature. Perturbations in the economic equilibrium occur frequently enough. In the absence of compensatory action, it is not certain whether the equilibrium will be regained spontaneously. And fluctuations in the economic activity with large amplitudes may not be desirable from a social point of view. Also, the fluctuations may not be permissible from political and other angles.

Knowledge about the functioning of the economy and the effects of instruments is far from perfect. In addition, there are elements of a random nature that influence the economic system. As a result, policy decisions have to be made in the presence of uncertainty. Principal objective of control theory is to improve system performance, particularly when uncertainty is involved. The regulator situation of control theory, where a quadratic penalty is put on the deviations of the system state from the desired state, is eminently suited for the analogous short-term economic policy problem. In addition, however, the use of control actions need to be limited in practice. A quadratic penalty on the use of control actions can be imposed by generalizing the performance criterion. The economic regulation problem can thus be formulated as a linear stochastic problem with a quadratic criterion of optimum performance.

The closed-loop control policy leads to a simple method of determining optimal control actions, given appropriate statistics based on the available information. The determination of these statistics, namely, the conditional mean and the error covariance matrix of the system state, can take place separately. The relationship between the system state and the information data is kept in view explicitly. The observations of the economic activity are supposed to contain observation errors, and changing or fragmentary economic indicators can be included.

The incorporation of information is achieved through the Kalman filter. This filter provides minimum-variance, unbiased estimates of the system state, conditional on the available information. Since the filter updates the estimates in a recursive fashion, it has practical advantages. Past information need not be used again, since its effect is summarized in the earlier estimates, which are revised optimally in the light of the fresh data.

1.2. SOME RESULTS IN STOCHASTIC CONTROL THEORY

In Chapter 2 of the present study, a specific situation in stochastic control theory is considered. The system of interest is represented in it by a set of linear discrete-time equations. The transitions of the system state over time are affected by additive random disturbances, which may be considered as the contribution of effects not included in the system state, and of uncertain elements or 'noise'. Although the system state is of main concern, it is assumed, in the general representation, not directly measurable or observable. Only certain linear combinations of the system state variables are observed, and the observation mechanism is furthermore imperfect in that observation errors are introduced. In the mathematical description, the errors are represented by additive random disturbances. The 'noise' processes considered are uncorrelated having zero mean. The system state can be manipulated by use of control variables, that also enter the model in a linear fashion. In an optimal control situation, it is required that the control actions be employed so as to optimize some measure or criterion of system performance. The objective of the chosen control problem is to minimize the expectation of a quadratic function that penalizes the deviations of the system state from some 'desired' state and the use of control actions over a finite horizon of operation.

The strategy of, what is technically called, the closed-loop control policy is adopted in order to obtain the optimal control actions. In such a policy, the information consisting of previous control action values and system observations available upto the moment when control action is to take place is used in specifying the control actions. The analytical derivation of the closed-loop policy results in a useful and convenient solution of the stochastic control problem. It is found that determination of optimal control actions can be achieved in two separate steps. In one, the estimation of the system state based on available information is obtained, and subsequently, the optimal control action is determined from a deterministic system, which is derived from the stochastic system by replacing the random quantities by their conditional expectations. It thus emerges that the Certainty Equivalence Principle, which was developed in the context of optimization problems in operations research and economics (cf. Simon (1956), Theil (1964)), holds also for the stochastic control situation considered.

With the passage of time, more and more information accumulates. In order to apply the Certainty Equivalence Principle, appropriate conditional expectations need to be provided. The Kalman filter (Kalman (1960), Kalman and Bucy (1961)) is a convenient technique to revise optimally the estimates based on past information in the light of fresh information alone. In effect, the conditional mean and the error covariance of the system state summarize all the accumulated information, so far as the optimal control situation is concerned. The Kalman filter enables their updating recursively. Similarly, the past conditional statistics can be revised recursively with the coming of later information. The results of Meditch (1967) facilitate this.

The optimal control actions resulting from the closed-loop policy depend not only on the latest observation but also on the past observations. The policy attaches appropriate weights to the sequence of observations. The outcome is that the optimal control action is a weighted average of latest and past observed errors in prediction. This feature is technically called the 'proportional-plus-integral control', and it results in smoother control actions and adds to overall stability.

1.3. ECONOMIC REGULATION AS CLOSED-LOOP CONTROL

Closed-loop control has been defined as the policy where controls are 'some deterministic function of the current and past observation on the system state variable and of past employed controls' (Aoki (1967), page 11). In present practice, economic regulatory policies tend to be based only on current rather than current and past observations of the state of the economy, and they do not take into account the fact that later more decisions will have to be taken in the light of new information. But in general they fit closed-loop control situations very well: there is a 'noisy' system, which is imperfectly observed, and the observations are used to derive policies that are intended to improve the situation, measured by some 'performance index'.

In the regulation case, the performance index, or objective function, can be the weighted sum of squares of differences between the predicted values of certain target variables and their ideal, or desired values. This is in line with the quadratic preference functions popularized by Theil (1964).

The controls are the 'instruments' as defined by Tinbergen (1956) and the model of the system can be a Tinbergen linear policy model too.

The major difference with the Tinbergen and Theil approaches lies in the accumulation of observations of the state of the system. While the usual economic literature assumes the current state of the system completely known, so that all uncertainty is concentrated in the future, the closed-loop control technique allows for the systematic treatment of the much more realistic case when information is scarce, contradictory and inexact. It is here that 'filter' techniques come in to make the most of the information available.

The mathematical analysis of linear closed-loop control with a quadratic performance index leads to such important results as separability of prediction (assuming unchanged policies) from policy determination and certainty equivalence, even if the poor quality of information is taken into account. This formally justifies Tinbergen's similar separation, and his concentration on deterministic analysis.

The Netherlands may well be a uniquely suitable testing ground for refined economic regulation techniques, as the government is able to vary tax rates (within bounds) without prior approval of the Parliament. This facility was introduced specifically for purposes of regulation, and the interest of the Netherlands policy-makers in short-term regulation also follows from the fact stated by Abert (1969) that they tend to have a quarterly discussion of the state of the economy. The experiments related in this study refer to quarterly control situations in the Netherlands in 1961–1970.

The linear model of the Netherlands employed here is almost entirely aprioristic. Estimation of model coefficients is not discussed in this study at all.

1.4. EXPERIMENTS WITH CLOSED-LOOP ECONOMIC REGULATION

The Kalman apparatus requires not only a linear model of the functioning of the economy, but also a model of the observation sub-system that converts the system state vector into published statistics, an objective function, and covariance matrices of the noises existing in the system. All of these are provided in a more or less intuitive fashion. Predictions of the exogenous variables are made available, obtained by applying Kalman filter methods to their statistical observations.

A typical 'experiment' is then to predict the target variables two, three and four quarters ahead, and to determine optimal values for the three instruments two and three quarters ahead. In the 40 quarters of the 1961–1970 period, 36 of these 'experiments' are possible. They have been performed assuming that all of 10 quarterly statistical series would be available, and the effect of leaving out one of these 10 at a time has been observed for each of the 10 series.

2. Control of linear stochastic systems

2.1. INTRODUCTION

In this chapter, a variety of results from the formal theory of control will be represented, on the basis of which the determination of quantitative macroeconomic policy will be demonstrated in the rest of the volume. Since economic phenomena are as a rule random in nature, the theory pertaining to control of *stochastic* systems is relevant. Linear control theory is used because linear analysis is usually assumed sufficient for short-term policy considerations (see, for instance, Goldberger (1959), page 20, Tinbergen (1956), page 55). The advantage of linear analysis is, of course, its simplicity and generality.

It is assumed that the process of interest is sufficiently accurately described by a linear dynamic mathematical model in which additive uncorrelated random disturbances appear. The observations or measurements of the system state are also considered disturbed by an additive independent random process.

The objective of a control policy is to specify control actions such that an index or criterion of system performance is optimized. The criterion chosen here is the statistical expectation of a quadratic function which reflects short-term economic policy aims. The strategy of what is technically known as the closed-loop control policy, in which the history of observations and the past control actions is taken into account, is adopted in order to obtain optimal control actions. A significant outcome is that the Certainty Equivalence Principle is applicable to the situation under consideration so that the value of control actions can be determined from a corresponding deterministic control problem in which the expected values appear in place of the respective random quantities. The solution to the stochastic control problem then consists of, first, estimating the mean of the system state and, subsequently, specifying control action values. This separation of estimation and control specification greatly simplifies obtaining the solution to the overall control problem.

The estimation of the conditional mean of the system state is based on the history of observations and past control actions. As time progresses, the set of historical information grows with the arrival of fresh observations. An estimation method that needs all the historical information in order to arrive at the updated conditional mean would be rather inconvenient and inefficient. We will employ the Kalman filter which overcomes this shortcoming by recursively updating the past statistics in the light of fresh information. The estimated conditional mean and covariance of the system state based on the historical information form a set of 'sufficient statistics' for the solution of the stochastic control problem in the sense that they summarize all the relevant information contained in the historical data for specifying completely the control actions. The Kalman filter updates these sufficient statistics when fresh data are available and thereby obviates the need to use the past data again. Great economy in computations is thus achieved. The estimates furnished by Kalman filter have additional desirable statistical properties. They are the unbiased linear estimators of the system state that have minimum variance for any positive semi-definite objective function.

The closed-loop control policy we use results in a feedback effect which weights successive statistical information data appropriately and thus yields relatively smooth control actions. This 'proportional-plus-integral' property resembles 'exponential smoothing'. Two further sections refer to subjects which are important in later chapters. Smoothing will be used in Chapter 6 while stability and controllability are of importance in Chapters 5, 6 and 7.

The filter must exhibit stability in the sense that a set of finite data should never result in unbounded estimates. Otherwise, the outcome could not be used for control purposes. The stability of the Kalman filter is therefore considered for a time-invariant dynamic system. When certain conditions are fulfilled, the conditional error covariance matrix of the estimates approaches a unique finite steady-state value. We shall use this property in this study.

These are the issues dealt with in the following sections. The description is based largely on Aoki (1967), Kalman (1960), Kalman and Bucy (1961) and Meditch (1967). Other sources are referred to in the text.

2.2. A FINITE HORIZON STOCHASTIC CONTROL PROBLEM

Consider a stochastic system which is represented mathematically by the following set of linear difference equations:

$$x_{i+1} = A_i x_i + B_i u_i + C_i \xi_i, \qquad i = 0, 1, 2, \ldots, N-1 \tag{2.2.1}$$

$$y_{i+1} = H_{i+1} x_{i+1} + \eta_{i+1} \tag{2.2.2}$$

where

i: 0, 1, 2,..., N; discrete time index
N: fixed finite time-horizon
x_i: an n-dimensional system state vector at time i
u_i: an m-dimensional control action vector at time i
y_i: an l-dimensional observation vector at time i
ξ_i: a q-dimensional vector of random disturbances at time i
η_i: an l-dimensional vector of random disturbances affecting the observations at time i
A_i, B_i, C_i, H_i: given non-random matrices of appropriate dimensions at time i.

The random disturbances ξ_i and η_i are supposed to have zero mean and to be independent of each other at all times and uncorrelated amongst themselves over different time periods. In addition, they are assumed to have finite covariance matrices. In symbols, for $i, j = 0, 1, 2, \ldots, N$,

$$E(\xi_i) = 0 \tag{2.2.3}$$

$$E(\eta_i) = 0 \tag{2.2.4}$$

$$E(\xi_i \xi_j^T) = \begin{cases} Q_i, & \text{when} \quad i = j \\ 0, & \text{otherwise} \end{cases} \tag{2.2.5}$$

$$E(\eta_i \eta_j^T) = \begin{cases} R_i, & \text{when} \quad i = j \\ 0, & \text{otherwise} \end{cases} \tag{2.2.6}$$

$$E(\xi_i \eta_j^T) = 0 \tag{2.2.7}$$

In Equations (2.2.5) and (2.2.6), the covariance matrices Q_i and R_i are positive-(semi) definite matrices. At some point in the discussion, we will make the additional assumption that the random processes ξ_i and η_i are gaussian (normally distributed). For such processes, the knowledge of the means and covariances is sufficient for a complete description of the probability distributions. In engineering, such a gaussian sequence which is uncorrelated over time is known as 'white noise' (Box and Jenkins, (1970)).

We assume that the probability density function of the system state at $i = 0$, $p(x_0)$, is known. If, in addition, $p(x_0)$ is assumed to be normally distributed, the process x_i becomes a gaussian process, given 'white noise'.

This representation of a linear stochastic system is fairly general. It is also typical or standard in control theory. The time-varying matrices A_i, B_i and C_i take into account the changes in the system characteristics with time. The time-varying feature implies different transitions of the system state at different times (the system is 'non-stationary'). The presence of the random disturbances ξ_i in Equation (2.2.1) implies that these transitions of the system state are known in a probabilistic sense only. The presence of the observation Equation (2.2.2) recognizes the empirical fact that the system state may not be observed as such; only certain linear combinations, perturbed by random disturbances are measured as observation variables y_i.

The sequence of control actions $u_0, u_1, \ldots, u_{N-1}$ affects the system over time. In an optimal control problem, it is required to decide on the control sequence such that some chosen criterion related to system performance is maximized or minimized. We adopt a fairly general quadratic performance criterion. The objective will be to determine the control sequence which minimizes the statistical expectation of the quadratic function. The objective is

$$\min_{u_0, u_1, \ldots, u_{N-1}} E(J)$$

where

$$J = \sum_{i=1}^{N} (x_i^T V_i x_i + u_{i-1}^T W_{i-1} u_{i-1}) \tag{2.2.8}$$

and V_i, W_i being given non-random positive-(semi) definite matrices. Let

$$G_i(x_i, u_{i-1}) = x_i^T V_i x_i + u_{i-1}^T W_{i-1} u_{i-1} \tag{2.2.9}$$

then the objective can be expressed as

$$\min_{u_0, u_1, \ldots, u_{N-1}} E(J) = \min_{u_0, u_1, \ldots, u_{N-1}} E\left[\sum_{i=1}^{N} (x_i^T V_i x_i + u_{i-1}^T W_{i-1} u_{i-1})\right]$$

$$= \min_{u_0, u_1, \ldots, u_{N-1}} E\left[\sum_{i=1}^{N} G_i(x_i, u_{i-1})\right] \tag{2.2.10}$$

2.3. DERIVATION OF THE CLOSED-LOOP CONTROL POLICY

In a closed-loop control policy (a term employed in control engineering), the control action u_i at time i is determined on the basis of the past and current observations $\{y_0, y_1, \ldots, y_i\}$ and the past control sequence $\{u_0, u_1, \ldots, u_{i-1}\}$. For the system under consideration, such a policy can be derived in a recursive manner, starting with the case when $i = N-1$. As optimality is maintained throughout, the resulting u_0 will be optimal also.

For convenience we will adopt the notation y^i for the sequence $\{y_0, y_1, \ldots, y_i\}$, the notation u^i for the sequence $\{u_0, u_1, \ldots, u_i\}$, and similarly for others. In addition, the arguments of a function will sometimes be omitted so long as there is no possibility of causing ambiguity.

a. *The last stage*

The last stage of decision-making is reached when the time index $i = N-1$. Only the last control action u_{N-1} remains to be specified in this situation. The performance criterion $E(J)$ is minimized by minimizing $E(G_N(x_N, u_{N-1}))$ with respect to u_{N-1}. From the definition of conditional expectations, we have

$$E(G_N) = \int [E(G_N \mid y^{N-1}, u^{N-2})]\, p(y^{N-1}, u^{N-2})\, d(y^{N-1}, u^{N-2}) \quad (2.3.1)$$

The quantity $E(G_N)$ is minimized if the conditional expectation $E(G_N \mid y^{N-1}, u^{N-2})$ is minimized for every y^{N-1} and u^{N-2}. This is a consequence of the quadratic nature of G_N. In the last stage, given the assumption that y^{N-1} and u^{N-2} are observed or known, the last control action u_{N-1} will actually be determined and thus will be non-random. All the previous control actions u_i's are non-random as well.

For convenience, we let λ_N stand for the conditional expectation:

$$\lambda_N = E(G_N \mid y^{N-1}, u^{N-2}) \quad (2.3.2)$$

or

$$\lambda_N = \int (G_N(x_N, u_{N-1}))\, p(x_N \mid y^{N-1}, u^{N-2})\, dx_N \quad (2.3.3)$$

The probability density function can be rewritten as follows:

$$p(x_N \mid y^{N-1}, u^{N-2}) =$$

$$\int p(x_N \mid x_{N-1}, u_{N-1}, \xi_{N-1})\, p(x_{N-1}, u_{N-1}, \xi_{N-1} \mid y^{N-1}, u^{N-2}) \times$$

$$\times\, d(x_{N-1}, u_{N-1}, \xi_{N-1}) \quad (2.3.4)$$

The random disturbances ξ_i are by assumption statistically independent of other processes. Hence

$$p(x_{N-1}, u_{N-1}, \xi_{N-1} \mid y^{N-1}, u^{N-2}) = \\ = p(x_{N-1}, u_{N-1} \mid y^{N-1}, u^{N-2}),\, p(\xi_{N-1}) \qquad (2.3.5)$$

The quadratic function appearing in (2.3.3) is

$$G_N(x_N, u_{N-1}) = x_N^T V_N x_N + u_{N-1}^T W_{N-1} u_{N-1} \qquad (2.3.6)$$

In this function, we can substitute

$$x_N = A_{N-1} x_{N-1} + B_{N-1} u_{N-1} + C_{N-1} \xi_{N-1} \qquad (2.3.7)$$

If the expectation is taken, the terms linear in ξ_{N-1} disappear because the ξ_{N-1} are independent of all other variables. The quadratic term in ξ_{N-1} is

$$E(\xi_{N-1}^T C_{N-1}^T V_N C_{N-1} \xi_{N-1}) = \operatorname{trace}(V_N C_{N-1} Q_{N-1} C_{N-1}^T) \qquad (2.3.8)$$

After substitutions of (2.3.5) and (2.3.8) in (2.3.3),

$$\lambda_N = \int [x_{N-1}^T (A_{N-1}^T V_N A_{N-1})\, x_{N-1} + 2u_{N-1}^T (B_{N-1}^T V_N A_{N-1}) \cdot x_{N-1} \\ + u_{N-1}^T (W_{N-1} + B_{N-1}^T V_N B_{N-1})\, u_{N-1}] \cdot p(x_{N-1}, u_{N-1} \mid y^{N-1}, \\ u^{N-2})\, d(x_{N-1}, u_{N-1}) + \operatorname{trace}(V_N C_{N-1} Q_{N-1} C_{N-1}^T) \qquad (2.3.9)$$

Let u_{N-1}^* represent the optimal u_{N-1} which minimizes λ_N and γ_N^* the corresponding minimum. That is

$$\gamma_N^* = \min_{u_{N-1}} \lambda_N \qquad (2.3.10)$$

The optimal control action u_{N-1}^* is found by minimizing the expression on the right hand side of Equation (2.3.9). Denote by x_{N-1}^* the conditional expectation of x_{N-1} given y^{N-1} and u^{N-2}, or,

$$x_{N-1}^* = E(x_{N-1} \mid y^{N-1}, u^{N-2}) \qquad (2.3.11)$$

The optimal control action u_{N-1} is found to be

$$u_{N-1}^* = -[W_{N-1} + B_{N-1}^T V_N B_{N-1}]^{-1} [B_{N-1}^T V_N A_{N-1}]\, x_{N-1}^* \qquad (2.3.12)$$

if the inverse exists. If the inverse does not exist, it should be read as a pseudo-inverse. In the sequel all such inverses will be assumed to exist.

Adopt the following symbols:

$$\Lambda_{N-1} = [W_{N-1} + B_{N-1}^T V_N B_{N-1}]^{-1} [B_{N-1}^T V_N A_{N-1}] \qquad (2.3.13)$$

$$\Pi_1 = (A_{N-1}^T V_N B_{N-1}) \cdot \Lambda_{N-1} \tag{2.3.14}$$

$$\Gamma_1 = A_{N-1}^T V_N A_{N-1} - \Pi_1 \tag{2.3.15}$$

$$\sigma_1 = tr(V_N C_{N-1} Q_{N-1} C_{N-1}^T) + \tag{2.3.16}$$
$$+ E[(x_{N-1} - x_{N-1}^*)\ \Pi_1\ (x_{N-1} - x_{N-1}^*) \mid y^{N-1}, u^{N-2}]$$

Then the optimal value of the control action and the function λ_N can be expressed as

$$u_{N-1}^* = -\Lambda_{N-1} x_{N-1}^* \tag{2.3.17}$$

and

$$\gamma_N^* = \min_{u_{N-1}} \lambda_N = E[(x_{N-1}^T \Gamma_1 x_{N-1} + \sigma_1) \mid y^{N-1}, u^{N-2}] \tag{2.3.18}$$

as can be found by substitution of u_{N-1}^* into (2.3.9).

b. *Last two stages*

Consider the situation when $i = N-2$ so that only the last two control actions u_{N-2} and u_{N-1} remain to be determined. Assume that the past observation and control sequences y^{N-2} and u^{N-3} are known. The control u_{N-2} affects both G_{N-1} and G_N.

It is seen that

$$E[G_{N-1}(x_{N-1}, u_{N-2}) + G_N(x_N, u_{N-1})] =$$
$$= \int [E((G_{N-1} + G_N) \mid y^{N-2}, u^{N-3})]\ p(y^{N-2}, u^{N-3})\ d(y^{N-2}, u^{N-3}) \tag{2.3.19}$$

Because the information y^{N-2}, u^{N-3} is contained in y^{N-1}, u^{N-2},

$$E(G_N \mid y^{N-2}, u^{N-3}) =$$
$$= \int E(G_N \mid y^{N-1}, u^{N-2})\ p(y_{N-1}, u_{N-2} \mid y^{N-2}, u^{N-3})\ d(y_{N-1}, u_{N-2})$$
$$= \int \lambda_N p(y_{N-1}, u_{N-2} \mid y^{N-2}, u^{N-3})\ d(y_{N-1}, u_{N-2}) \tag{2.3.20}$$

The optimal policy requires that the minimum of λ_N will be sought whatever values y_{N-1} and u_{N-2} may take, therefore λ_N can be replaced by its optimal value γ_N^*.

We now define

$$\gamma_{N-1} = E(G_{N-1} \mid y^{N-2}, u^{N-3}) + \int \gamma_{NP}^{*} p(y_{N-1} \mid y^{N-2}, u^{N-3})\, dy_{N-1} \tag{2.3.21}$$

Analogous to Equation (2.3.3) we define

$$\lambda_{N-1} = E(G_{N-1} \mid y^{N-2}, u^{N-3}) \tag{2.3.22}$$

and rewriting the probability density function similarly to (2.3.4) we find

$$\lambda_{N-1} = \int G_{N-1}(x_{N-1}, u_{N-2})\, p(x_{N-1} \mid x_{N-2}, u_{N-2}, \xi_{N-2}) \cdot$$
$$\cdot p(x_{N-2}, u_{N-2} \mid y^{N-2}, u^{N-3}) \cdot p(\xi_{N-2})\, d(x_{N-1}, x_{N-2}, u_{N-2}, \xi_{N-2}) \tag{2.3.23}$$

Equation (2.3.21) can then be rewritten as

$$\gamma_{N-1} = \lambda_{N-1} + \int \gamma_{NP}^{*} p(y_{N-1} \mid y^{N-2}, u^{N-3})\, dy_{N-1} \tag{2.3.24}$$

Making use of Equation (2.3.18), we obtain

$$\int \gamma_{NP}^{*} p(y_{N-1} \mid y^{N-2}, u^{N-3})\, dy_{N-1} =$$
$$= E[(x_{N-1}^{T} \Gamma_1 x_{N-1} + \sigma_1) \mid y^{N-2}, u^{N-3}] \tag{2.3.25}$$

Substitution in Equation (2.3.24) yields

$$\begin{aligned} \gamma_{N-1} &= E[(x_{N-1}^{T} V_{N-1} x_{N-1} + u_{N-2}^{T} W_{N-2} u_{N-2}) \mid y^{N-2}, u^{N-3}] + \\ &\quad + E[(x_{N-1}^{T} \Gamma_1 x_{N-1} + \sigma_1) \mid y^{N-2}, u^{N-3}] = \\ &= E[(x_{N-1}^{T} (V_{N-1} + \Gamma_1)\, x_{N-1} + \\ &\quad + u_{N-2}^{T} W_{N-2} u_{N-2} + \sigma_1) \mid y^{N-2}, u^{N-3}] \end{aligned} \tag{2.3.26}$$

As in the case of the last stage situation, the optimal control action u_{N-2}^{*} which minimizes the conditional expectation γ_{N-1} and the optimal γ_{N-1}^{*} are determined by the following relationships:

$$u_{N-2}^{*} = -\Lambda_{N-2} x_{N-2}^{*} \tag{2.3.27}$$

$$\gamma_{N-1}^{*} = \min_{u_{N-2}} \gamma_{N-1} =$$
$$= E[(x_{N-2}^{T} \Gamma_2 x_{N-2} + \sigma_2) \mid y^{N-2}, u^{N-3}] \tag{2.3.28}$$

where

$$x^*_{N-2}=E(x_{N-2}\mid y^{N-2},u^{N-3}) \tag{2.3.29}$$

$$\Lambda_{N-2}=[W_{N-2}+B^T_{N-2}(V_{N-1}+\Gamma_1)\,B_{N-2}]^{-1}\times \\ \times[B^T_{N-2}(V_{N-1}+\Gamma_1)\,A_{N-2}] \tag{2.3.30}$$

$$\Pi_2=A^T_{N-2}(V_{N-1}+\Gamma_1)\,B_{N-2}\Lambda_{N-2} \tag{2.3.31}$$

$$\Gamma_2=A^T_{N-2}(V_{N-1}+\Gamma_1)\,A_{N-2}-\Pi_2 \tag{2.3.32}$$

and

$$\sigma_2=\sigma_1+tr[(V_{N-1}+\Gamma_1)\,C_{N-2}Q_{N-2}C^T_{N-2}]+ \\ +E[(x_{N-2}-x^*_{N-2})^T\Pi_2(x_{N-2}-x^*_{N-2})\mid y^{N-2},u^{N-3}] \tag{2.3.33}$$

c. *General case*

When $i=k$, the controls $u_k, u_{k+1},\ldots, u_{N-1}$ need to be determined such that the sum $E(\sum_{k+1}^N G_i)$ is minimized. At this stage the policy is based on the observation sequence y^k and the past control actions u^{k-1} which are assumed known. When y^k and u^{k-1} are actually observed, the appropriate control action u_k will actually be determined. Thus, we need to consider the optimization of the criterion conditioned on the observation and control sequences y^k and u^{k-1}. Symbolically,

$$E\left(\sum_{k+1}^{N} G_i\right)=\int\left[E\sum_{k+1}^{N}(G_i\mid y^k,u^{k-1})\right]p(y^k,u^{k-1})\,d(y^k,u^{k-1}) \tag{2.3.34}$$

We adopt the notation

$$\gamma^*_{k+1}=\min_{u_k,\ldots,u_{N-1}}\left[E(G_{k+1}\mid y^k,u^{k-1})+\sum_{k+2}^{N}E(G_i\mid y^k,u^{k-1})\right] \tag{2.3.35}$$

The principle of optimality from the theory of dynamic programming is now invoked:

'An optimal policy has the property that whatever the initial state and decision are, the remaining decisions must constitute an optimal policy with regard to the state resulting from the first decision' (Bellman, 1957).

This principle can be applied in our case because at every stage a state

x_k exists which results from earlier control actions $u_0, \ldots, u_{k-1}$ and the disturbances ξ_i, $i=0, \ldots, k-1$ but which cannot be affected by future control actions $u_k, \ldots, u_{N-1}$.

We therefore have

$$\gamma_{k+1}^* = \min_{u_k} \left[E(G_{k+1} \mid y^k, u^{k-1}) + \int \gamma_{k+2}^* p(y_{k+1} \mid y^k, u^{k-1})\, dy_{k+1} \right] \tag{2.3.36}$$

Letting

$$\lambda_{k+1} = E(G_{k+1} \mid y^k, u^{k-1}) \tag{2.3.37}$$

Equation (2.3.36) becomes

$$\gamma_{k+1}^* = \min_{u_k} \left[\lambda_{k+1} + \int \gamma_{k+2}^* p(y_{k+1} \mid y^k, u^{k-1})\, dy_{k+1} \right] \tag{2.3.38}$$

Equation (2.3.38) is valid for all $k=0, 1, 2, \ldots, N-1$ if it is assumed that

$$\gamma_{N+1}^* = 0 \tag{2.3.39}$$

because for $k=N-1$ the proper expectation for optimization is

$$\gamma_N^* = \min_{u_{N-1}} \lambda_N$$

General recursive relationships for the determination of the optimal control actions $u_i^* = u_k^*$ are now found to be, analogous to the previous two cases, as follows:

$$u_i^* = -\Lambda_i x_i^* \qquad 0 \le i \le N-1 \tag{2.3.40}$$

where

$$x_i^* = E(x_i \mid y^i, u^{i-1}) \tag{2.3.41}$$

$$\Lambda_i = [W_i + B_i^T (V_{i+1} + \Gamma_{N-i-1}) B_i]^{-1} [B_i (V_{i+1} + \Gamma_{N-i-1}) A_i] \tag{2.3.42}$$

$$\Pi_{N-i} = [A_i^T (V_{i+1} + \Gamma_{N-i-1}) B_i] \Lambda_i \tag{2.3.43}$$

$$\Gamma_{N-i} = [A_i^T (V_{i+1} + \Gamma_{N-i-1}) B_i] - \Pi_{N-i} \tag{2.3.44}$$

The optimal value of the performance index at this stage given the knowledge of y^i and u^{i-1} is

$$\gamma_{i+1}^* = E[(x_i^T \Gamma_{N-i} x_i + \sigma_{N-i}) \mid y^i, u^{i-1}] \tag{2.3.45}$$

where

$$\sigma_{N-i} = \sigma_{N-i-1} + tr[(V_{i+1} + \Gamma_{N-i-1})\, C_i Q_i C_i^T] + \\ + E[(x_i - x_i^*)^T \Pi_{N-i}(x_i - x_i^*) \mid y^i, u^{i-1}] \qquad (2.3.46)$$

At time i given that the observation and control sequences y^i and u^{i-1} are known, the optimal control action u_i^* is non-random. It can be determined using the recursive relationships equations (2.3.42)–(2.3.46). This involves computation of the conditional expectation $x_i^* = E(x_i \mid y^i, u^{i-1})$. Knowledge of the conditional mean x_i^* enables one to find the complete solution to the optimal control problem. As the control actions turn out to be non-random and for convenience, they are omitted as conditioning arguments in the sequel.

2.4. SEPARATION OF SYSTEM STATE ESTIMATION AND CONTROL ACTION DETERMINATION

The quantities Λ_i, Γ_i and Π_i, which appear in equations (2.3.42)–(2.3.46) that determine the optimal control action, are independent of the random disturbances ξ_i and η_i, the observations y^i and the past control actions u^{i-1}. They are non-random quantities, and can be calculated before the control begins. The random processes ξ_i and η_i affect the conditional means $x_i^* = E(x_i \mid y^i)$ and the quantity σ_i. Since, as mentioned above, the optimal control actions u_i are completely specified when the means $E(x_i \mid y^i)$ are known, the optimal control policy consists of two separate phases:

1. estimation of the conditional mean x_i^* of the state x_i based on a set of observation data y^i and the past control actions u^{i-1}.
2. determination of the quantity $-\Lambda_i$ that gives the control action as a linear transformation of the estimated state. The linear transformation Λ_i, called the 'feedback coefficient' in engineering, is identical to the transformation that would be applied if there were no disturbances and if the system state had the estimated expected value (the conditional mean). In other words, the control actions can be determined from a corresponding non-random system in which the random quantities are replaced by their expected values.

This implies that the Certainty Equivalence Principle (cf. Simon (1956), Theil (1964)) which states that under certain conditions the optimal control policies for a stochastic system are identical to those for the related deterministic system in which expected values are substituted for the corresponding random quantities, is also applicable to the present system.

The separation of the (statistical) estimation phase and the (deterministic) optimization phase in the solution of the overall stochastic control problem is significant both from a theoretical and a practical viewpoint.

The execution of the optimization by means of a recursive system is another practical advance over solving the problem simultaneously for all stages.

2.5. RECURSIVE ESTIMATION OF THE SYSTEM STATE

The closed-loop control policy requires computation of the conditional expectation $E(x_i \mid y^i)$ in order to specify the optimal control action u_i^*. The observation y_i pertains to the interval i. It turns out that all the information contained in the past observation sequence y^{i-1}, i.e., $\{y_0, y_1, y_2, \ldots, y_{i-1}\}$, which is relevant for the control policy, can be summarized in the conditional expectation $E(x_{i-1} \mid y^{i-1})$. It will now be shown that the desired conditional mean $E(x_i \mid y^i)$ can be obtained from the preceding estimate and its covariance matrix, and the fresh observation y_i only. The whole history of information data $\{y_0, y_1, \ldots, y_{i-1}\}$ need not thus be used again.

The conditional mean and covariance matrix can be updated by means of the Kalman filter. This filter is also referred to as Kalman-Bucy filter or Kalman-Wiener filter (cf. Aoki (1967), Deutsch (1965)). We shall indicate the filter in broad outlines, and only for the problem in hand.

The conditional mean $E(x_i \mid y^i)$ can be found if the whole conditional probability density function $p(x_i \mid y^i)$ is known. Making use of Bayes' Rule, this function can be determined recursively, starting from the initial function $p(x_0 \mid y_0)$.

For convenience, we assume that $p(x_0)$ is a normal probability density function with mean μ_0 and covariance Σ_0. Symbolically,

$$p(x_0) = N(\mu_0, \Sigma_0) \tag{2.5.1}$$

We also assume that the disturbances ξ_i and η_i are normally distributed. We can now derive the moments of the conditional probability density function $p(x_0 \mid y_0)$ which is also normal.

The following theorem will be applied:

Let $\begin{bmatrix} x_0 \\ y_0 \end{bmatrix}$ be normally distributed with mean μ and covariance matrix Σ.

Let μ and Σ be partitioned into

$$\begin{bmatrix} \mu_x \\ \mu_y \end{bmatrix} \quad \text{and} \quad \begin{bmatrix} \Sigma_{xx} & \Sigma_{xy} \\ \Sigma_{yx} & \Sigma_{yy} \end{bmatrix}$$

Then the conditional probability function of x_0 given y_0 is

$$p(x_0 \mid y_0) = N[\mu_x + \Sigma_{xy}\Sigma_{yy}^{-1}(y_0 - \mu_y), \Sigma_{xx} - \Sigma_{xy}\Sigma_{yy}^{-1}\Sigma_{yx}] \tag{2.5.2}$$

(cf. for instance Dhrymes (1970), page 19).

Now

$$y_0 = H_0 x_0 + \eta_0 \tag{2.5.3}$$

hence

$$E(y_0) = H_0\mu_0, \operatorname{cov}(y_0) = H_0\Sigma_0 H_0^T + R_0, E(x_0 y_0^T) = \Sigma_0 H_0^T$$

Define

$$\hat{x}_0 = \mu_0 \tag{2.5.4}$$

$$P_0 = \Sigma_0 \tag{2.5.5}$$

$$K_0 = P_0 H_0^T (H_0 P_0 H_0^T + R_0)^{-1} \tag{2.5.6}$$

$$x_0^* = \hat{x}_0 + K_0[y_0 - H_0\hat{x}_0] \tag{2.5.7}$$

which is the conditional mean given in the above theorem. Furthermore, let

$$S_0 = (I - K_0 H_0) P_0 \tag{2.5.8}$$

then S_0 is the covariance of the conditional distribution. Putting it directly,

$$p(x_0 \mid y_0) = N(x_0^*, S_0) \tag{2.5.9}$$

Now assume that for some i it has been proved that

$$p(x_i \mid y^i) = N(x_i^*, S_i) \tag{2.5.10}$$

In a similar manner we can then prove that Equation (2.5.10) holds for $i+1$ and as it already applies to $i=0$ (Equation (2.5.9)) we have proved that it is valid for all $i=0, 1, 2, \ldots$.

The x_{i+1}^* and S_{i+1} are recursively defined by the formulae

$$P_{i+1} = A_i S_i A_i^T + C_i Q_i C_i^T \tag{2.5.11}$$

$$K_{i+1} = P_{i+1} H_{i+1}^T (H_{i+1} P_{i+1} H_{i+1}^T + R_{i+1})^{-1} \tag{2.5.12}$$

$$S_{i+1} = (I - K_{i+1} H_{i+1}) P_{i+1} \tag{2.5.13}$$

$$\hat{x}_{i+1} = A_i x_i^* + B_i u_i \tag{2.5.14}$$

$$x_{i+1}^* = \hat{x}_{i+1} + K_{i+1}[y_{i+1} - H_{i+1}\hat{x}_{i+1}] \tag{2.5.15}$$

It is seen from Equations (2.5.11)–(2.5.15) that the conditional mean x_{i+1}^* and the conditional covariance S_{i+1} are computable from their corresponding previous values, namely x_i^* and S_i, and the new observation y_{i+1}. This indicates that x_i^* and S_i summarize all the relevant information contained in the historical data y^i; that is, they form a set of sufficient statistics for y^i for a complete solution to the stochastic control problem.

Equations (2.5.11)–(2.5.15) comprise the Kalman filter algorithm. The term 'filter' for this kind of algorithm arose in communications engineering where signals are to be separated from noise. The algorithm furnishes Bayesian estimates x_i^* as linear combinations of the set of observation data. The quantity x_i^* is the orthogonal projection of x_i on the Hilbert space of observations y^i (cf. Kalman (1960), Kalman and Bucy (1961), Deutsch (1965), Luenberger (1969)). It is a linear, minimum-variance, unbiased estimator of a system state over a sample of observation data y^i. This means that it is 'optimal for *any* positive-semidefinite quadratic optimality criterion' (Luenberger (1969), page 91). The optimality of the Kalman prediction-correction scheme is also arrived at, although from a different starting point, by Krasnakevich and Haddad (1969).

Apart from the computational convenience residing in the recursive nature of the algorithm, its greatest merit lies in this optimality property. Regardless of the objective of the control policy, in the problem stated estimation of the existing state can best be done by means of the Kalman filter.

2.6. PROPORTIONAL-PLUS-INTEGRAL CHARACTER OF FEEDBACK CONTROL POLICY

We shall now examine in what manner successive observations of the system state are taken into account in specifying optimal control actions.

From Equation (2.3.40) the optimal control action at time $i+1$ is

$$u_{i+1}^* = -\Lambda_{i+1} x_{i+1}^* \tag{2.6.1}$$

In the Kalman filter algorithm the conditional mean x_{i+1}^* is determined by the relation

$$x_{i+1}^* = \hat{x}_{i+1} + K_{i+1}[y_{i+1} - H_{i+1}\hat{x}_{i+1}] \tag{2.6.2}$$

The quantity within brackets in Equation (2.6.2) is the error in prediction based on information available up to one interval before. Let $\hat{e}_i(1)$ denote the one interval ahead prediction error at time i, i.e.,

$$\hat{e}_i(1) = y_i - H_i\hat{x}_i \tag{2.6.3}$$

Because the one interval ahead conditional prediction of the state vector is given by

$$\hat{x}_{i+1} = A_i x_i^* + B_i u_i^* \tag{2.6.4}$$

Equation (2.6.2) can be written as

$$x_{i+1}^* = A_i x_i^* + B_i u_i^* + K_{i+1}\hat{e}_{i+1}(1) \tag{2.6.5}$$

Substituting the proper optimal control action in Equation (2.6.5) yields

$$x_{i+1}^* = K_{i+1}\hat{e}_{i+1}(1) + (A_i - B_i\Lambda_i)\, x_i^* \tag{2.6.6}$$

Recursion and substitution leads to

$$x_{i+1}^* = K_{i+1}\hat{e}_{i+1}(1) + (A_i - B_i\Lambda_i)\,\{K_i\hat{e}_i(1) + (A_{i-1} - B_{i-1}\Lambda_{i-1})\, x_{i-1}^*\} \tag{2.6.7}$$

Successive recursion and substitution, and after setting $\hat{x}_0 = 0$ for convenience only, finally gives

$$x_{i+1}^* = K_{i+1}\hat{e}_{i+1}(1) + \sum_{j=0}^{i}\left[\prod_{k=j}^{i}(A_k - B_k\Lambda_k)\right] K_j\hat{e}_j(1) \tag{2.6.8}$$

The optimal control action at time $(i+1)$ is then

$$u_{i+1}^* = -\Lambda_{i+1}K_{i+1}\hat{e}_{i+1}(1) - \Lambda_{i+1}\sum_{j=0}^{i}\left[\prod_{k=j}^{i}(A_k - B_k\Lambda_k)\right] K_j\hat{e}_j(1) \tag{2.6.9}$$

The first term on the right hand side of Equation (2.6.9) represents an action which is proportional to the current one interval ahead prediction error while the second term denotes an action proportional to a weighted sum of past errors in prediction.

A similar relationship can be derived between the control actions u_i and the individual observation vectors y_i. Rearrange Equation (2.6.2) in the form

$$x_{i+1}^* = K_{i+1}y_{i+1} + [I - K_{i+1}H_{i+1}]\,\hat{x}_{i+1} \tag{2.6.10}$$

Substitution of Equations (2.6.4) and (2.6.1) in this relation leads to

$$x_{i+1}^* = K_{i+1}y_{i+1} + (I - K_{i+1}H_{i+1})\,(A_i - B_i\Lambda_i)\, x_i^* \tag{2.6.11}$$

Successive recursion finally results in (again after setting $\hat{x}_0 = 0$)

$$x^*_{i+1} = K_{i+1} y_{i+1} + \sum_{j=0}^{i} \left\{ \prod_{k=j+1}^{i+1} (I - K_k H_k)(A_{k-1} - B_{k-1} \Lambda_{k-1}) \right\} K_j y_j \tag{2.6.12}$$

Therefore, the optimal control action at time $(i+1)$ is

$$u^*_{i+1} = -\Lambda_{i+1} K_{i+1} y_{i+1} - \Lambda_{i+1} \times$$
$$\times \sum_{j=0}^{i} \left\{ \prod_{k=j+1}^{i+1} (I - K_k H_k)(A_{k-1} - B_{k-1} \Lambda_{k-1}) \right\} K_j y_j \tag{2.6.13}$$

It is again noted that the optimal control action is composed of parts proportional to the present observation and to a weighted sum of past observations.

In control engineering, such a scheme is referred to as proportional-plus-integral control.

The proportional-plus-integral nature of the feedback control policy results from the minimization of the expectation of a quadratic function. Because the control actions are linear functions of some weighted average of current and past prediction errors, they will not react so sharply to recent prediction errors as a policy based on those recent errors alone. Thus, when prediction errors are caused largely by uncorrelated random exogenous disturbances, the appropriate control action can be smoother than it would otherwise be. This adds to general continuity and stability.

The effect of distant past observations on the control action u^*_{i+1} will diminish provided that the weights $\prod_{k=j+1}^{i+1} (I - K_k H_k)(A_{k-1} - B_{k-1} \Lambda_{k-1})$ are smaller than K_{i+1} in some sense.

The proportional-plus-integral feature resembles the concept of exponentially weighted moving averages (cf. Holt *et al.* (1960)). The latter is well-known in literature on time series analysis and forecasting. It is also called by the name 'exponential smoothing' (Brown (1963)). It has been reported by the author (1970a) that an optimal scheme of exponential smoothing for the special case of a single time series with only a trend, as developed by Theil and Wage (1964) (see also Nerlove and Wage (1964)) can be shown to be a particular case of the Kalman filter. The present framework of the linear non-stationary stochastic control problem is of course much more general and takes care of not only prediction but also policy-making in a multivariable setting with, in general, inaccurate measurements, possibly fewer in number than the system state variables.

The quantity $\hat{x}_{i+1}$ is calculated in Equation (2.5.14). It is in fact a conditional expectation, one interval ahead, of x_{i+1} based on the observations y^i, because

$$p(x_{i+1} \mid y^i) = p((A_i x_i + B_i u_i + C_i \xi_i) \mid y^i) \tag{2.7.1}$$

and therefore

$$p(x_{i+1} \mid y^i) = N[A_i x_i^* + B_i u_i,\ A_i S_i A_i^T + C_i Q_i C_i^T] \tag{2.7.2}$$

i.e.,

$$p(x_{i+1} \mid y^i) = N[\hat{x}_{i+1}, P_{i+1}] \tag{2.7.3}$$

In other words, $\hat{x}_{i+1}$ is a prediction of x_{i+1} based on y^i.

In general, given a sequence of observations $y^j = \{y_0, y_1, \ldots, y_j\}$, three different types of estimates can be distinguished. Let $\mu_{i/j}$ denote an estimate of x_i based on y^j. Then

1. when $i > j$, $\mu_{i/j}$ is the prediction of x_i based on y^j,
2. when $i = j$, $\mu_{i/j}$ is called the filtered estimate of x_i given y^i,
3. when $i < j$, $\mu_{i/j}$ is called the smoothed estimate of x_i based on y^j.

We will adopt the following notation for some particular estimates

$$\hat{x}_{i+1} = \mu_{i+1/i}, \text{ for the one-period prediction}$$

$$x_i^* = \mu_{i/i}, \text{ for the filtered estimate}$$

$$x_{i/j}^* = \mu_{i/j},\ i < j, \text{ for the smoothed estimate}$$

The Kalman filter furnishes predictions $\hat{x}_{i+1}$ and the filtered estimates x_i^*. All the past estimates $x_{i/j}^*$ can be revised whenever fresh observations become available because they provide information not only on the current state but also on all past states, as a result of the dynamic nature of the system. An algorithm developed by Meditch (1967) for obtaining the linear, minimum-variance, unbiased smoothed estimates of x_i given observations y^j, $i < j$, is as follows:

$$x_{i/j}^* = x_i^* + J_i[x_{i+1}^* - \hat{x}_{i+1}], \qquad i = j-1, j-2, \ldots, 0 \tag{2.7.4}$$

where

$$J_i = S_i A_{i+1}^T P_{i+1}^{-1} \tag{2.7.5}$$

The quantities appearing in these relations are the same as defined earlier. The corresponding smoothing error covariance matrix is governed by

$$\mathrm{cov}(x^*_{i/j}) = S^*_{i/j} = S_i + J_i[S^*_{i+1/j} - P_{i+1}]\,J_i^T,\ i = j-1, j-2, \ldots, 0 \quad (2.7.6)$$

The Meditch algorithm is an estimate-correction scheme, like the Kalman filter.

2.8. CONTROLLABILITY, OBSERVABILITY AND STATIONARY STATISTICS

The conditional mean and covariance, as mentioned earlier, form a set of sufficient statistics, in the sense that they summarize relevant information contained in observation data, for the solution of the stochastic control problem. The system we consider is in general of the non-stationary variety. This means that, in particular, the mean and covariance of the process change with time. The corresponding conditional statistics are likewise non-stationary. It is important that they remain bounded in order to obtain proper solutions of the control problem. The Kalman filter is employed to provide these statistics in a recursive manner. Therefore, we wish that the filter be stable in the sense that a set of finite data should not result in unbounded estimates of the conditional mean and covariance. Kalman and Bucy (1961) report on the stability properties of the filter. Here we discuss a particular type of stability requirement in the context of a stationary specification of the stochastic system. The application to our non-stationary system will follow later.

Consider a system specification consisting of the following relations:

$$x_{i+1} = Ax_i + Bu_i + C\xi_i \quad (2.8.1)$$

$$y_{i+1} = Hx_{i+1} + \eta_{i+1} \quad (2.8.2)$$

The random processes ξ_i and η_i are assumed to have the following expectations:

$$E(\xi_i) = 0 \quad (2.8.3)$$

$$E(\eta_i) = 0 \quad (2.8.4)$$

$$E(\xi_i\xi_j^T) = \begin{cases} Q, & \text{when } i=j \\ 0, & \text{otherwise} \end{cases} \quad (2.8.5)$$

$$E(\eta_i\eta_j^T) = \begin{cases} R, & \text{when } i=j \\ 0, & \text{otherwise} \end{cases} \quad (2.8.6)$$

$$E(\xi_i\eta_j^T) = 0 \quad (2.8.7)$$

The present specification differs from that in Equations (2.2.1) and (2.2.2) in that the matrices A, B, C and H are given constant matrices and the 'noise' sequences ξ_i and η_i are stationary processes, because Q and R are also given and constant.

The concepts of controllability and observability are now introduced because they will be required for the discussion of filter stability.

Controllability. A linear system of the type in Equations (2.8.1) and (2.8.2) is called completely controllable if at any instant of time $i>0$, the corresponding state vector x_i can be transformed to the original state x_0 by means of a suitable control function (Deutsch (1965)).

In the filtering context, controllability with respect to the random disturbances ξ_i need be considered. In a constant linear system, a necessary and sufficient condition for complete controllability (with respect to ξ_i) is

$$\text{row rank}\,[C \;\; (AC)\ldots \;\; (A^{n-1}C)]=n \tag{2.8.8}$$

where n is the dimension of the state vector and $C, (AC), \ldots, (A^{n-1}C)$ denote the columns of an (n, nq)-dimensional matrix, q being the dimension of vector ξ_i.

Observability. A linear system of the type in Equations (2.8.1) and (2.8.2) is called completely observable if the covariance matrix $S_i=\text{cov}(x_i \mid y^i)$ associated with the conditional probability density function $p(x_i \mid y^i)$ remains bounded positive-definite as $i\to\infty$ (Aoki (1967)).

Observability is of importance when one wishes to ensure that all the components of the state vector will be optimally estimated. A physical interpretation of complete observability is that all the modes of system response become available to the observation mechanism in a finite time.

For the constant linear system in Equations (2.8.1) and (2.8.2), a necessary and sufficient condition for complete observability is that

$$\text{row rank}\,[H^T \;\; (HA)^T\ldots \;\; (HA^{n-1})^T]=n \tag{2.8.9}$$

where, as before, n denotes the length of the vector x_i and $H^T, (HA)^T, \ldots, (HA^{n-1})^T$ stand for the columns of the $(n, l\cdot n)$-dimensional matrix.

We now indicate an important theorem of Kalman and Bucy (1961). Its conclusions will be used in Chapters 5 and 6.

Given, for the linear stochastic system of Equations (2.8.1) and (2.8.2), that

1. the matrices A, B, C and H are finite, time-independent matrices,
2. the 'noise' covariance matrices Q and R are finite, positive-definite, and time-invariant,

3. the linear system is completely controllable (with respect to ξ_i),
4. the linear system is completely observable.

Then, it follows

i. the Kalman filter is uniformly asymptotically stable,
ii. every solution S_i of the recurrence relationships Equations (2.5.11), (2.5.12) and (2.5.13) which results from initializing with any symmetric non-negative matrix S_0 tends uniformly to a unique finite positive-definite matrix $\bar{S}$.

In our computations, we will use these conditional covariance matrices $\bar{S}$ for filtering and smoothing.

It should be noted that the error covariance matrix S_i does not vanish; it approaches a finite positive-definite matrix. This is in contrast with what happens in some other estimation methods, like the least squares method, in which the error covariance of the estimates of the regression coefficients tends to zero asymptotically. The finite error covariance resulting from the filter implies that the true values of the state vector can never be learned from the observations y_i. This is in accordance with the nature of the stochastic process we are dealing with.

3. Economic regulation based on frequent statistical information

3.1. ECONOMIC REGULATION IN LITERATURE

In economic literature, the Tinbergen (1956) book on 'Economic Policy: Principles and Design' provides a convenient starting point. Tinbergen defines a 'policy model' which predicts the effect of 'instruments' on 'targets'. In all of his policy models, except the very simplest, several instruments act on several targets simultaneously: there is no one-to-one correspondence between targets and instruments.

Most of the models discussed are related to problems of regulation, that is, of trying to bring the economy to a preferred state, sometimes called 'equilibrium' or a 'stable' state. This concept appears to be only loosely connected to the 'general economic equilibrium' of classical economics (cf., for instance, Brems (1968)) or to the 'stability' of dynamic systems in the Lyapunov sense (cf. Section 5.7).

Examples given (pp. 62 and 98) of features of a desired state are: a low and constant rate of unemployment, prices that do not diverge too much from world prices, a zero deficit in the balance of payments.

Once a reduction in deviations from a desired state is taken as the main aim of regulatory policy it is natural to formalize this aim in an objective function consisting of a weighted sum of squares of these deviations. Probably the first more or less practical example is in a paper by Van den Bogaard and Theil (1959), quoted and extended in Theil (1964).

If the policy model is linear, and the objective function quadratic, additive stochastic disturbances can be replaced by their expected means for the purpose of determining the optimum policy. This is the Certainty Equivalence Principle, which after its first formulation by Simon (1956) was generalized by Theil, and applied ever since because of its obvious great advantages.

Applications of classical control theory to economic regulation problems may be said to have started with Phillips' (1954) paper on 'Stabilization Policy in a Closed Economy', although Tustin (1952) made some attempts before him. Closed-loop control of the proportional-cum-integral-cum-derivative type is treated there. It is, however, essentially non-stochastic, and by the time Allen (1967) reviewed the matter this was still the case. Fox, Sengupta and Thorbecke (1966) do not discuss stochastic control either in an otherwise wide-ranging survey.

In control theory, objective functions of several variables ('performance indices') had meanwhile been introduced. An early example of their application to regulation problems appears to be a paper by Kalman and Koepcke (1958). Then Kalman (1960) combined several developments in one article, which provides most of the theory used in the present study. Not only is closed-loop control of a stochastic system with a quadratic performance criterion discussed but linear filter theory is applied as well.

Filter theory is another branch of applied mathematics and statistics where quadratic functions are minimized. It lends itself readily to incorporation in stochastic control theory for linear systems with quadratic criteria.

One of the results of Kalman (1960) is certainty equivalence, now shown to exist even for a closed-loop policy that is a policy which depends on observations accumulated and on previous control actions. Another closely allied result is the separability of prediction and policy determination.

Little of this appears to have penetrated economic literature, which remains occupied mostly with problems of estimation, or stability, or optimal time-paths in deterministic cases.

3.2. CONTROL THEORY APPLIED TO ECONOMIC REGULATION

Wishart (1969) gives a schematic representation of a stochastic control system as described by Fel'dbaum (1953). This is shown in Figure 3.1, with designations fitting the regulation problem to be analyzed. Applied to the economic regulation problem, various elements can be interpreted in the following way.

The main system is the (national) economy, not including the subsystem to be described. Its state is observed by an observation subsystem, in particular a statistical agency, which produces observations y, adding 'observation noise' η. The observations enter into the filtering and control subsystem, typically the government agency which evaluates the information and converts it into predictions $\hat{x}$ of the system state x. Out of this comparison arises the advice for regulatory policy u, which is received by the execution subsystem, that is, the executive branch of the government.

At this function, a specific 'noise' ε comes in which distorts the policy action; the actual policy followed is then shown to influence the main system, which itself is subject to 'noise' ξ. As the observations y are taken into account for further regulatory action, Figure 3.1 represents a closed-loop control system in which there is an arrangement to measure the system state and to make the measurements available to the decision-

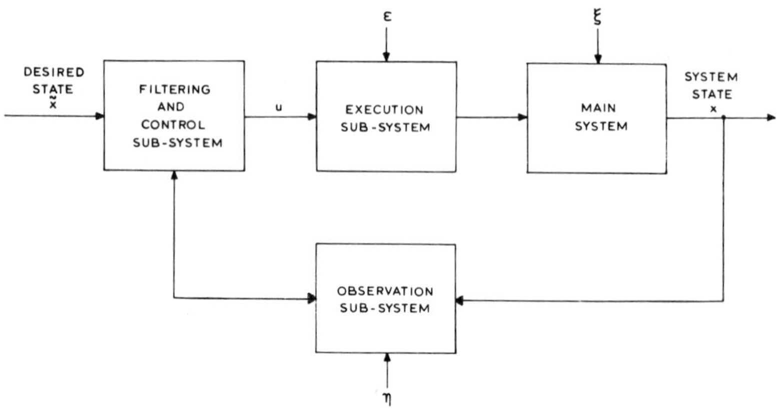

Figure 3.1. Schematic representation of a stochastic control system

making subsystem for incorporation into subsequent decisions. Thus, the decisions can be made not only in the light of the information which is already available on the past, and possibly on the current, states but also in the light of the fact that the system will be observed in the future also. The closed-loop arrangement is a device to reduce the losses caused by the uncertainties in the system.

The objective of regulation is to let the system approximate a desired state as closely as possible. In the economic context, this means that certain variables, such as unemployment, or prices, are watched and predicted, and their deviations from desired values are reduced as far as possible through appropriate controls. Such a desired value may be explained in various ways. It may, for instance, be held that rising prices are disadvantageous to recipients of fixed incomes while falling prices discourage investment. Fluctuating prices may lead to speculative behaviour and make physical production less efficient.

Following Tinbergen (1956), the means of economic policy can be distinguished as quantitative, qualitative, and reforms according to their nature. Reforms, such as employee participation in industrial management, are of the fundamental type, implying changes in the underlying social and cultural values of a community. Such changes are of necessity slow. Qualitative means of economic policy, for instance introducing built-in-flexibility to dampen cyclical swings, are also comparatively slow to put into effect. The quantitative means of economic policy are relatively easy to handle in the short-run, and are therefore suited to offset the fluctuations in the economic activity brought about by diverse perturbations. The quantitative means employed for economic regulation are termed 'instruments'.

The specification of quantitative policy is made complicated because of interdependence amongst economic phenomena. In general, the instruments of policy have conflicting effects from the point of view of regulation. If production is to be raised in order to increase employment, prices are likely to rise at the same time. Similarly, measures taken to regulate, say, the balance-of-payments possibly have adverse effect on employment. The instruments not only have short-run conflicting effects, but may also prove to be inconsistent from the point of view of their effects in the long-run. For example, credit creation for short-run regulation may give rise to inflation in future if suitable counter measures are not taken. To consider another example, if investment is restricted with a view to reducing pressure on prices, capacity for production in the future is affected. The use of a mathematical model is advisable to avoid inconsistency in use of instruments with respect to the desired objectives.

In a mathematical representation, it is possible to incorporate not only such basic requirements as the fact that the domestic production and balance of foreign trade equal domestic consumption, investment and inventory formation, but also relevant behaviour relationships such as demand functions.

In Chapter 2, a mathematical representation of a stochastic control system is given. In it, the random disturbances ε appearing in Figure 3.1 are not treated separately. The model of the system is linear. The quadratic performance criterion J in Section 2.2 is appropriate for regulation inasmuch as the deviations from the desired state can be minimized. It also allows penalties to be imposed on the use of instruments so that they are kept within reasonable limits.

In Tinbergen's (1956, p. 52) discussion of economic regulation, the random disturbance processes ε, ξ, and η are ignored, and the logic of decision-making is considered in a deterministic framework. He notes, however, that this is not always admissible. Simon (1956) and Theil (cf. his book (1964)) show that under their assumptions the Certainty Equivalence Principle is applicable so that the decisions for the stochastic problem can be obtained from a corresponding deterministic problem in which expectations conditional on latest information of random quantities appear. Thus, after conditional expectations have been calculated, it is allowed to ignore the random disturbances, which are assumed to be zero mean processes.

For the stochastic control problem of Chapter 2, the Certainty Equivalence Principle holds. Thus, in the present application too, the decisions can be made from a deterministic model, replacing random quantities by their conditional expectations. This result is more general than that of Theil. In the model considered in Chapter 2, the role played by information data is explicitly kept in view and the closed-loop control strategy, in which both the present and future decisions take into account all pertinent information, is adhered to.

The systematic approach to determine the values of instruments for economic regulation, as proposed by Tinbergen (1956), can be executed in a number of phases. Because the execution of the policy takes time and the information about the economic activity is available with some delay only, forecasts of the economic activity or state are needed. Predictions are made assuming no changes in the instrument values but taking into account the likely developments in the exogenous variables. When the predictions, based on the latest information, are found to be different from the desired activity, changes in the policy instruments are indicated.

In Section 7.5, it is seen that the overall control strategy for the determination of regulatory policy consists of the following steps:

Step 1. Updating the conditional expectations in the light of fresh observations.

Step 2. Prediction of the future system state under assumption of no policy change.

Step 3. Calculation of the weighted deviations of the predicted states, under assumption of no policy change, from the desired future states.

Step 4. Determination of the optimal instrument values.

It is noted that the prediction of the future course of the economy is separated from the phase of specifying optimal instrument values. The approach of Tinbergen (1956) is thus vindicated.

In this study, the Kalman filter is employed to obtain the conditional expectations recursively.

Finally, as described in Chapter 6, the same 'filter' technique to accumulate the evidence of successive observations is sufficient for the extrapolation of exogenous variables, which are required in the prediction and control of future economic activity.

3.3. INFORMATION PROCESSING FOR ECONOMIC REGULATION

Most governments prepare their budgets, their reviews of the state of the nation and of the world at large, and their statements on forthcoming policy, on an annual basis. In those countries where governments systematically attempt to regulate the economy, predictions are drawn up annually, even if they are sometimes made by means of quarterly forecasting models.

However, it is also found that governments do prepare 'interim budgets' and use other means to regulate the economy outside the conventional annual rhythm. These interpolated policies generally arise out of disturbing statistical information (about the balance-of-payments deficit, say) that has appeared after the regular, annual prediction has been digested.

The processing of such interim information into predictions sufficiently clear-cut to persuade the government to take remedial action is necessarily less careful and detailed than the leisurely annual predictions routine, and therefore bound to be unduly alarmist at times, with the result that many governments will tend to ignore interim reports much more easily than they tend to do in the case of the customary annual reviews.

The loss caused by less than optimal processing of recent statistical information can be expected to be relatively great because such information is known to be incomplete, conflicting and inexact to a degree far surpassing that of the carefully sifted and analyzed statistical data going into annual surveys. It is, for instance, quite possible to see unemployment increase in the same month that the index of industrial production and the volume of retail sales both rise strongly. On some important phenomena, such as changes in stocks of final products, hardly any short-term information may be available. Statistics may be just plainly wrong as Morgenstern (1950) pointed out at some length.

In the Netherlands, the need for an optimal quarterly information processing system is perhaps greater than anywhere else. The Netherlands government has persuaded Parliament to let it change taxes at short notice without prior parliamentary approval (the so-called 'wobble tax' law), thus enabling the government to shorten the lag between the arrival of information and the execution of policy related to it.

Abert (1969) states that the Netherlands Central Planning Bureau prepares quarterly reports which are not published but circulated among policy-making agencies. These reports are additional to the annual survey published in the beginning of the year under the title 'Central Economic Plan'. One of them is the basis of the 'Macro-Economic Outlook' memorandum which appears with the budget in September. The Netherlands may therefore well be the country where optimal filtering techniques can be used with greatest effect.

A striking feature of the formulation in Chapter 2 is that it explicitly allows the possibility of changing sets of observation data from period to period; and the fact that in practice, a lesser number of economic indicators are available than the number of components of the system state with which one wishes to characterize the economic activity in any period. The Kalman filter optimally updates the system state estimates and predictions in the light of changing data of varying abundance. The annual prediction exercises mentioned above solve this problem differently: they are based on more statistical information, and gaps and discrepancies are removed by careful *ad hoc* estimations. For the quarterly predictions, a linear filter may well be the best technique.

3.4. SELECTION OF A POLICY MODEL

The macro-economic process is vast and complex. It contains elements of natural, technological, institutional, international, and psychological character. Even in the simple model used in the present study, some 30 variables are distinguished to represent the economy, and 50 odd time series of observations over the 40 quarters of the period 1961–1970 are made use of in the computations. Consequently, describing an economy in the form of a mathematical model is a demanding endeavour.

Because controlled experiments with economic systems are not possible, model building in economics must follow a radically different approach from that in situations of physical sciences and engineering where such experiments are possible. A common resort is to establish hypotheses of economic behavior. A model in the form of a set of mathematical equations describes in an approximate manner the behaviour of various groups of economic agents. Economic theory tries to establish hypotheses and derive conclusions from them.

As it purports to explain empirically occurring phenomena, an economic theory must be related to the real world. Inadequacy in this respect sometimes leads to severe pronouncements. Shoeffler (1955), for instance, categorically states that economics is not a nomothetic, empirical science at all but only an art of forecasting and decision making.

Available theories of economic behaviour at times exhibit considerable divergence amongst themselves. Recourse can be had to statistical method in order to choose from competing hypotheses. Statistical testing itself is conditional in nature and can never ascertain a causal explanation to be 'true' or 'false' (Tinbergen (1956)). Besides statistical significance, physical or empirical significance is equally important. However, *ex post* statistically significant relations are considered superior to completely aprioristic models even if they are less relevant *a priori* to *ex ante* situations.

The model used in the present study, while not entirely unrelated to statistical estimation by various econometricians, is largely aprioristic in nature. It is believed that the conclusions to be drawn from the study are not affected by this fact. Those who are interested in the estimation of coefficients of a quarterly model for the Netherlands may consult Driehuis (1972).

4. A simple quarterly model for the Netherlands economy

4.1. INTRODUCTION

The model presented below is largely based on the common elements of earlier models for the Netherlands; it does not contain any new ideas at all about the way the Netherlands economic system works. It will be used for experiments in the sixties, but it is bound to fit statistical observations in the sixties less well than the recent quarterly model estimated by Driehuis (1972) from 1951–1965 data, and it will probably be less accurate in the seventies too. Its main merit is simplicity: it represents the minimum of relationships that economic analysts would wish to see incorporated in a short-term policy model.

It is a quarterly model because a time-period of a quarter is neither too long nor too short for a realistic representation of possible changes in economic policy in response to recent information.

The present chapter provides all the elements required to construct the state-space representation part of the model, that is the part that describes transitions of states over time.

4.2. SALIENT FEATURES

The mathematical representation of the Netherlands economy described here is in the spirit of the econometric model contained in the 1955 Central Economic Plan of the Central Planning Bureau (1955) in the sense that various parameters were given values on largely *a priori* grounds. The choice of economic variables is influenced by the econometric models developed by the Central Planning Bureau (1961), Van den Beld (1968) and Driehuis (1972). There are, however, significant differences – mostly simplifying – from these in many respects. The monetary variables, like the interest rate and the money supply, are ignored. The scheme of lagged influences is much simpler than in the Driehuis specification. Unlike the latter quarterly econometric model, the presentation here explicitly accounts for seasonal effects to describe the seasonal nature of the economy when considered on a quarterly basis. The variables are kept in their absolute measurement units without transformation to logarithms or percent changes. Later on, however, first-order differences from given reference values are involved.

In the model, the stochastic nature of the phenomena is explicitly kept in view by the presence of exogenous random disturbance processes. The random disturbances affecting the endogenous variables are distinguished from the possible errors in observations. The latter are related to the observation mechanism considered in the next chapter (Chapter 5). The disturbance processes involved here are meant to represent not only the approximations in modelling of the economic activities, i.e., the effect of phenomena excluded, but also to account for numerous incidental causes which are primarily stochastic in nature.

The twenty endogenous variables are determined by a set of as many equations. The first ten of them are what are technically known as the 'reaction' or 'behaviour' equations, in view of the fact that they describe the adaptation behaviour of economic entities to changing causes. The remaining ten are definition equations or identities. All of the reaction equations are linear but the definitions that link value, quantity and price variables introduce non-linearity into the model, which is later removed by linearization.

4.3. LIST OF VARIABLES AND UNITS OF MEASUREMENTS

In the following list of variables, economic and other phenomena are distinguished as endogenous, viz. the variables governed by the mathematical model of the economy, exogenous, i.e., phenomena that affect the endogenous variables but the generation of which is not explained within the model, and control variables, those which are at the discretion of the policy-makers who by their use influence the economic activity characterized by the endogenous variables. In literature, control variables are synonymously known as decision-variables, policy variables or instruments. The exogenous and control variables are sometimes grouped together under the label 'data'. The past values of the endogenous variables are also sometimes included in data variables.

The trend and seasonal variables are endogenous in some respects, as they are generated by the model. On the other hand, they are not affected by the instruments and exogenous variables, and may therefore be considered as exogenous to the restricted system. They are estimated and predicted at the same time as the endogenous variables.

The mathematical model is intended to represent the quarterly macroeconomic phenomena in the Netherlands. Although the flow variables in it refer to quarters of a year, they are measured at annual rates. In contrast, the stock variables are measured as averages over each quarter.

Prices are measured as index numbers to the base 1963 = 1 because this required least conversion.

In order to simplify the notation, certain conventions are adopted for the symbols which are employed to represent diverse phenomena. The time index t, which designates a quarter t, is understood implicit in each symbol, i.e., a symbol α for a variable in quarter t would be more completely represented by α_t. The lagged variable α_{t-k} is simply denoted by

Table 4.1. Endogenous variables

No.	*Symbol*	*Phenomenon*	*Units*
1	P	Dependent labour supply (work force)	100,000 persons
2	a	Number of wage earners	100,000 persons
3	C	Total private consumption at current prices	bln.gld.* at current prices
4	i	Gross private investment in fixed assets at constant prices	bln.gld. at 1963 prices
5	n	Increase in stocks	—— do ——
6	mg	Import of goods at fixed prices	—— do ——
7	l	Wage rate of wage earners	10,000 gld. per person
8	pc	Price level of private consumption	index no., 1963 at 1
9	pi	Price level of private investments	—— do ——
10	TZ	Tax on non-wage income	bln. gld.
11	v	Gross national expenditure at fixed prices	bln. gld. at 1963 prices
12	U	Number of unemployed persons	100,000 persons
13	c	Quantity of total private consumption in fixed prices	bln. gld. at 1963 prices
14	I	Value of gross private investments at current prices	bln. gld. at current prices
15	L	Wage income	—— do ——
16	Z	Non-wage income	—— do ——
17	Y	Gross national product at current prices	—— do ——
18	LD	Disposable wage income	—— do ——
19	ZD	Disposable non-wage income	—— do ——
20	M	Value of import of goods and services at current prices	—— do ——

* 'bln. gld.' stands for 1,000,000,000 guilders. Similarly, 'mln. gld.' implies 1,000,000 guilders.

α_{-k}. In the expenditure categories, capital letters refer to values in current prices, lower case letters to volumes and prices. An underline distinguishes the symbols for control and exogenous variables from those for the endogenous phenomena. The exogenous time-trend and seasonal variation are designated by a prefix *d* or *s*, respectively, to the symbol for the corresponding endogenous variable; thus $d\alpha$ and $s\alpha$. Similarly, random disturbances are denoted by a prefix ξ to the symbol for the phenomenon in the equation of which they first appear, for example, ξP. Finally, certain extraneously determined values are designated by a bar above the corresponding symbol of a variable, thus $\bar{\alpha}$. The use of these 'reference values' will appear from Sections 4.4 and 4.7.

Table 4.2. Exogenous variables

No.	*Symbols*	*Phenomenon*	*Units*
1	$\underline{az}$	Number of self-employed persons	100,000 persons
2	$\underline{g}$	Quantity of Govt. expenditure at constant prices	bln. gld. at 1963 prices
3	$\underline{G}$	Value of Govt. expenditure at current prices	bln. gld. at current prices
4	$\underline{b}$	Quantity of exports of goods at constant prices	bln. gld. at 1963 prices
5	$\underline{B}$	Value of export of goods at current prices	bln. gld. at current prices
6	$\underline{MD}$	Net import of services	bln. gld. at current prices
7	$\underline{pmg}$	Price of import of goods	index no., 1 in 1963

*Table 4.3. Control variables**

No.	*Symbol*	*Phenomenon*	*Units*
1	$\underline{TL}$	Tax on wages	bln. gld.
2	$\underline{TK}$	Indirect taxes	bln. gld.
3	$\underline{TRL}$	Change in the level of tax on wages	bln. gld.
4	$\underline{TRZ}$	Change in the level of tax on non-wage income	bln. gld.
5	$\underline{TRK}$	Change in the level of indirect taxes	bln. gld.

* Only three controls will be used as the following two relationships apply:
$\underline{TL} = \underline{\overline{TL}} + \underline{TRL}$ and $\underline{TK} = \underline{\overline{TK}} + \underline{TRK}$

Dependent labour supply

The change in dependent work force $(P-P_{-1})$ is assumed caused by changes in unemployment, a trend effect (dP_{-1}) to account for demographic and other gradually changing factors, a seasonal effect (sP) that recognizes *inter alia* the seasonal appearance of school leavers, and a random disturbance process (ξP_{-1}). The mathematical relationship is

$$P=P_{-1}-0.5(U_{-1}-U_{-3})+dP_{-1}+sP+\xi P_{-1} \qquad (4.4.1)$$

The effect of unemployment changes is distributed over two quarters. The coefficient -0.5 means that increases in unemployment depress the supply of labour (cf. Van den Beld (1968), CPB[1] (1961)). Married women and school leavers in particular do not always register as unemployed, and are therefore statistically outside the labour force.

Number of wage earners

The principal determinant of the change in the number of wage earners $(a-a_{-1})$ is the average change in the gross national expenditure over the preceding two quarters, with an elasticity of 0.3. With trend added, the following equation governs the number of wage earners in the model.

$$a=a_{-1}+0.15\,\frac{\bar{a}}{\bar{v}}(v_{-1}-v_{-3})+da_{-1}+sa+\xi a_{-1} \qquad (4.4.2)$$

The symbols $\bar{a}$ and $\bar{v}$, as explained in Section 4.3, represent given reference values of a and v, respectively. In the model described in the 1961 Central Economic Plan (CPB (1961)), the corresponding elasticity of the variable a to changes in some reweighted measure of total sales, which reflects the volume of production, is 0.39.

Private consumption

The propensity to consume in a given quarter out of the disposable wage income (LD) of the same quarter is assumed to be 0.8. The effect of disposable non-wage income lags over the preceding two quarters

1. The abbreviation 'CPB' is employed for 'Central Planning Bureau'.

and the propensity is 0.4. The consumption equation employed is

$$C=0.8LD+0.2(ZD_{-1}+ZD_{-2})+dC_{-1}+sC+\xi C_{-1} \qquad (4.4.3)$$

The propensities were set *a priori* on the basis of experience with time-series analysis. In the model related to 1961 Central Economic Plan (cf. CPB (1961)), the corresponding figures were 0.86 and 0.41, respectively, on the basis of relationship among the value of *C*, *LD* and *ZD* in the year 1957.

Gross private investment in fixed assets

The investment equation determines gross private investment in real terms. The specific causal factor is the average over the preceding six quarters of the disposable non-wage income. The propensity is assumed to be 0.6, which fits with the propensity to consume of 0.4 although there is no need for these propensities to add up to exactly 1. The relationship is

$$i=0.1(ZD_{-1}+ZD_{-2}+ZD_{-3}+ZD_{-4}+ZD_{-5}+ZD_{-6})+ \\ +di_{-1}+si+\xi i_{-1} \qquad (4.4.4)$$

Increase in stocks

The increase in stocks (*n*) in a given quarter is determined by the average increase of the real gross national expenditure over the preceding two quarters. The relationship is

$$n=0.145(v_{-1}-v_{-3})+dn_{-1}+sn+\xi n_{-1}-0.5\xi mg_{-1} \qquad (4.4.5)$$

The coefficient 0.145 is derived from the coefficient of 0.29 used by the Central Planning Bureau in annual models (CPB (1961)). It is assumed that one-half of the disturbances in the relationship between imports and GNP (4.4.6) is reflected in stocks.

Imports of goods

The real gross national expenditure is supposed to exert the major influence on the imports of goods in real terms (*mg*). The effect has an elasticity of 1.5 and takes place with a lag of one quarter. The relationship is

$$mg=1.5\,\frac{\overline{mg}}{\bar{v}}\,v_{-1}+dmg_{-1}+smg+\xi mg_{-1} \qquad (4.4.6)$$

Wage rate per person

Changes in the wage rate are brought about by changes in the price of consumption goods, with a delay of one quarter, and also by the level of unemployment in the preceding quarter. The second causal factor recognizes the well-known Phillips curve (Phillips (1958)) in that the unemployment level shifts the wage rate. The elasticity of wage rate to consumption price changes is assumed to be 0.5. The unemployment effect enters the equation in a linear fashion. When the unemployment in a given quarter is higher by 1%, in the next quarter the wage rate increases by 0.4% less than the trend. Both coefficients are decidedly crude guesses. The equation is

$$l = l_{-1} + 0.5\,\frac{\bar{l}}{\bar{pc}}\,(pc_{-1} - pc_{-2}) - 0.04U_{-1} + dl_{-1} + sl + \xi l_{-1} \qquad (4.4.7)$$

Price level of private consumption

Wage rate (l), import prices ($\underline{pmg}$) and indirect taxes ($\underline{TK}$) determine the price of consumption (pc). The weights accorded to the wage rate and import prices are 70% and 30%, respectively. An increase in the amount of indirect taxes goes to increase the price of consumption by a factor $1/\bar{c}$, $\bar{c}$ being the volume of consumption. The equation is obtained as

$$\begin{aligned} pc = {} & pc_{-1} + 0.7(l - l_{-1}) + 0.3(\underline{pmg} - \underline{pmg}_{-1}) + \\ & + \frac{1}{\bar{c}}(\underline{TK} - \underline{TK}_{-1}) + dpc_{-1} + spc + \xi pc_{-1} \end{aligned} \qquad (4.4.8)$$

Price level of private investments

As with consumer goods, the major determinants of the price of investment goods are the wage rate and the price of imports, though the weights are different. The equation governing the change of the price of investments is

$$\begin{aligned} pi = {} & pi_{-1} + 0.4(l - l_{-1}) + 0.6(\underline{pmg} - \underline{pmg}_{-1}) + \\ & + dpi_{-1} + spi + \xi pi_{-1} \end{aligned} \qquad (4.4.9)$$

Tax on non-wage income

The principal factor determining the receipt of the tax is the average non-wage income over the preceding third, fourth, fifth and sixth quarters. An autonomous change can be brought about by the instrument variable $\underline{TRZ}$. The equation is

$$TZ = \frac{1}{4}\frac{\overline{TZ}}{\overline{Z}}(Z_{-3}+Z_{-4}+Z_{-5}+Z_{-6})+\underline{TRZ}+$$
$$+dTZ_{-1}+sTZ+\xi TZ_{-1} \qquad (4.4.10)$$

The coefficient $(\overline{TZ}/\overline{Z})$ represents the average tax rate. There are no similar relationships for the taxes on wage income and the indirect taxes because the corresponding average tax rates are much lower, so that less inaccuracy is introduced by leaving out the effect of fluctuations in wages or the national product.

4.5. IDENTITIES

In the model, there are ten endogenous variables which are determined from the identities implicit in the definition of the variables.

Volume of Gross National Expenditure

$$v=c+i+\underline{g}+n \qquad (4.5.1)$$

Number of Unemployed Persons

$$U=P-a-\underline{az} \qquad (4.5.2)$$

Quantity of Total Private Consumption

$$c=C/pc \qquad (4.5.3)$$

Value of Gross Private Investments

$$I=i\cdot pi \qquad (4.5.4)$$

Wage Income

$$L=a\cdot l \qquad (4.5.5)$$

Non-Wage Income

$$Z=Y-L-\underline{TK} \qquad (4.5.6)$$

Gross National Product

$$Y=C+I+\underline{G}+\underline{B}-M+n \qquad (4.5.7)$$

In this equation, the stock variable n appears in constant prices because the price effect on n is negligible.

Disposable Wage Income

$$LD = L - \underline{TL} \tag{4.5.8}$$

Disposable Non-Wage Income

$$ZD = Z - TZ \tag{4.5.9}$$

Imports of Goods and Services

$$M = \underline{pmg} \cdot mg + \underline{MD} \tag{4.5.10}$$

4.6. VARIABILITY OF THE RANDOM DISTURBANCES

In all the reaction equations in the model, Equations (4.4.1)–(4.4.10), random disturbances appear. As mentioned in Section 4.2, the random disturbances are introduced to account for errors caused by overlooking causal factors and to make the model reflect the uncertainty in the transition of the economy from quarter to quarter as represented by it. The random disturbances by themselves are all assumed uncorrelated with each other and over time. A correlation, however, is introduced between the random disturbances in the behaviour equations governing imports of goods and increase in stocks, Equations (4.4.6) and (4.4.5), by showing one half of the disturbance ξmg in the imports relationship entering into the stock increase behaviour. However, to repeat, before they are made part of any of the behaviour equations, the random disturbances are assumed to be uncorrelated among themselves and over time. In addition, they are assumed to have zero mean and fixed finite variances.

The variances of the disturbances were chosen almost entirely *a priori*. Table 4.4 (see p. 43) lists the standard-deviations in the units used for the corresponding endogenous variables, and also in percents of the levels of those variables in the year 1963.

4.7. TRANSFORMATION TO THE LINEAR REDUCED FORM

The mathematical model to represent the quarterly macro-economic activity in the Netherlands as proposed above is in a form known in

Table 4.4. Standard deviations of the random disturbances

	Random Disturbances	Standard Deviation (in units of corresponding variables)	Level of corresponding variables in 1963	Standard Deviation in percent of corresponding endogenous variable
1	ξP	0.45	44.220	1.0
2	ξa	0.35	35.360	1.0
3	ξC	0.89	31.662	2.8
4	ξi	0.78	9.900	7.8
5	ξn	0.71	0.480	14.8
6	ξmg	2.00	20.083	10.0
7	ξl	0.05	0.757	6.6
8	ξpc	0.02	1.000	2.0
9	ξpi	0.02	1.000	2.0
10	ξTZ	0.20	4.274	4.7

literature as the 'structural form'. The definitions involving value, quantity and price contain non-linearities, as noted earlier. In the structural form, the wage level l of a given quarter as determined by Equation (4.4.7) enters simultaneously in Equations (4.4.8) and (4.4.9) which determine price of consumption (pc) and investment (pi), respectively, of the same quarter. Such a non-linear structural form with simultaneous relationships is not readily amenable to the linear control theory described in Chapter 2. Suitable transformation, however, can result in a form to which linear control theory is applicable.

A linear approximation of the first order can be obtained by linearizing the model around a reference value or an 'operating point', as it is sometimes called in engineering, in the 20-dimensional space of the endogenous variables. Deviations of second and higher orders are then neglected.

Some care need be exercised in the selection of the reference values. If the set of reference values chosen does not satisfy the definition identities, that is, does not conform to the non-linearities in the model, the recursive filtering may lead to divergent results (Kau and Kumar (1969)). In the calculations reported in this volume, the values selected all satisfied the definition identities.

The reference values selected were equal to annual observations of the variables concerned and were used for four consecutive quarters of each calendar year.

In the equations used for the computations, most of the variables defined above were replaced by the sums of their reference values and the deviations from those reference values. For example, P was replaced by $\bar{P} + \Delta P$ (afterwards, $\bar{P}$ disappeared, but this does not apply to all reference values). The symbol Δ designates the difference between the value of a variable and its reference value; it does *not* indicate the difference with respect to a preceding value.

No such replacement was used for n, or for the trends and seasonal variables. As it represents stock changes, Δn can also be used for n.

After linearization and substitution of the identities the linearized 'reduced form' of the reaction equations is obtained as given below.

$$\Delta P=\Delta P_{-1}-0.5(\Delta U_{-1}-\Delta U_{-3})+dP_{-1}+sP+\xi P_{-1} \tag{4.7.1}$$

$$\Delta a=\Delta a_{-1}+0.15\frac{\bar{a}}{\bar{v}}(\Delta v_{-1}-\Delta v_{-3})+da_{-1}+sa+\xi a_{-1} \tag{4.7.2}$$

$$\begin{aligned}\Delta C=\;&0.8\bar{a}\left[\Delta l_{-1}+0.5\frac{\bar{l}}{\overline{pc}}(\Delta pc_{-1}-\Delta pc_{-2})-0.04\Delta U_{-1}\right]+\\&+0.8\bar{l}\left[\Delta a_{-1}+0.15\frac{\bar{a}}{\bar{v}}(\Delta v_{-1}-\Delta v_{-3})\right]+\\&+0.2[\Delta ZD_{-1}+\Delta ZD_{-2}]+\\&+(0.8\bar{a}dl_{-1}+0.8\bar{l}da_{-1}+dC_{-1})+(0.8\bar{a}sl+0.8\bar{l}sa+sC)+\\&+(0.8\bar{a}\xi l_{-1}+0.8\bar{l}\xi a_{-1}+\xi C_{-1})+\\&+[-0.8\underline{TL}]+\\&+[-\bar{C}-0.8\overline{LD}+0.8\bar{a}\cdot\bar{l}+0.4\overline{ZD}-0.032\bar{a}\cdot\bar{U}]\end{aligned} \tag{4.7.3}$$

$$\Delta i=0.1[\Delta ZD_{-1}+\cdots+\Delta ZD_{-6}]+[-\bar{i}+0.6\overline{ZD}]+di_{-1}+si+\xi i_{-1} \tag{4.7.4}$$

$$\Delta n=0.145(\Delta v_{-1}-\Delta v_{-3})+dn_{-1}+sn+\xi n_{-1} \tag{4.7.5}$$

$$\Delta mg=1.5\frac{\overline{mg}}{\bar{v}}\Delta v_{-1}+[0.5\overline{mg}]+dmg_{-1}+smg+\xi mg_{-1} \tag{4.7.6}$$

$$\begin{aligned}\Delta l=\;&\Delta l_{-1}+0.5\frac{\bar{l}}{\overline{pc}}(\Delta pc_{-1}-\Delta pc_{-2})-0.04\Delta U_{-1}+[-0.04\bar{U}]+\\&+dl_{-1}+sl+\xi l_{-1}\end{aligned} \tag{4.7.7}$$

$$\begin{aligned}\Delta pc=\;&\left(1+0.35\frac{\bar{l}}{\overline{pc}}\right)\Delta pc_{-1}-0.35\frac{\bar{l}}{\overline{pc}}\Delta pc_{-2}-0.028\Delta U_{-1}+\\&+\left[\frac{1}{\bar{c}}\underline{TRK}-\underline{TRK}_{-1}\right]+0.3(\underline{pmg}-\underline{pmg}_{-1})+[-0.028\bar{U}]+\end{aligned}$$

$$+(0.7dl_{-1}+dpc_{-1})+(0.7sl+spc)+(0.7\xi l_{-1}+\xi pc_{-1})$$

$$(4.7.8)$$

$$\Delta pi=\Delta pi_{-1}+0.2\frac{\bar{l}}{\overline{pc}}(\Delta pc_{-1}-\Delta pc_{-2})-0.016\Delta U_{-1}+$$

$$+[0.6(\underline{pmg}-\underline{pmg}_{-1})]+(0.4dl_{-1}+dpc_{-1})+$$

$$+0.4(sl+spi)+(0.4\xi l_{-1}+\xi pi_{-1})+[-0.016\bar{U}] \qquad (4.7.9)$$

$$\Delta TZ=0.25\frac{\overline{TZ}}{\bar{Z}}(\Delta Z_{-3}+\cdots+\Delta Z_{6})+[\underline{TRZ}]+$$

$$+dTZ_{-1}+sTZ+\xi TZ_{-1} \qquad (4.7.10)$$

The linearized form of identities is given below in Equations (4.7.11) through (4.7.20). In order to obtain the reduced form, the Equations (4.7.1) through (4.7.10) would have to be substituted in Equations (4.7.11) through (4.7.20). This is not attempted here because the substitution is more easily executed numerically on the computer.

$$\Delta v=\frac{1}{\overline{pc}}\Delta C-\frac{\bar{C}}{\overline{pc}^2}\Delta pc+\Delta i+\Delta n+[\underline{g}]+\left[\bar{v}+\bar{i}+\frac{\bar{C}}{\overline{pc}}\right] \qquad (4.7.11)$$

$$\Delta U=\Delta P-\Delta a+[-\underline{az}]+[-\bar{U}+\bar{P}-\bar{a}] \qquad (4.7.12)$$

$$\Delta c=\frac{1}{\overline{pc}}\Delta C-\frac{\bar{C}}{\overline{pc}}\Delta pc+\left[-\bar{c}+\frac{\bar{C}}{\overline{pc}}\right] \qquad (4.7.13)$$

$$\Delta I=\overline{pi}\cdot\Delta i+\bar{i}\cdot\Delta pi+[-\bar{I}+\bar{i}\cdot\overline{pi}] \qquad (4.7.14)$$

$$\Delta L=\bar{a}\cdot\Delta l+\bar{l}\cdot\Delta a+[-\bar{L}+\bar{a}\cdot\bar{l}] \qquad (4.7.15)$$

$$\Delta Z=\Delta C+\overline{pi}\cdot\Delta i+\bar{i}\cdot\Delta pi-\underline{pmg}\cdot\Delta mg+\Delta n-\bar{a}\cdot\Delta l-\bar{l}\cdot\Delta a-$$

$$-\underline{TK}+[+\underline{G}+\underline{B}-\underline{MD}-\underline{pmg}\cdot\overline{mg}]+$$

$$+[-\bar{Z}+\bar{C}+\overline{pi}\cdot\bar{i}-\bar{a}\cdot\bar{l}] \qquad (4.7.16)$$

$$\Delta Y=\Delta C+\overline{pi}\cdot\Delta i+\bar{i}\cdot\Delta pi-\Delta mg\cdot\underline{pmg}+\Delta n+$$

$$+[\underline{G}+\underline{B}-\underline{MD}-\underline{pmg}\cdot\overline{mg}]+[-\bar{Y}+\bar{C}+\overline{pi}\cdot\bar{i}] \qquad (4.7.17)$$

$$\Delta LD=\bar{a}\cdot\Delta l+\bar{l}\cdot\Delta a-\underline{TL}+[-\overline{LD}+\bar{a}\cdot\bar{l}] \qquad (4.7.18)$$

$$\Delta ZD=\Delta C+\overline{pi}\cdot\Delta i+\bar{i}\cdot\Delta pi-\underline{pmg}\cdot\Delta mg+\Delta n-\bar{a}\cdot\Delta l-\bar{l}\cdot\Delta a-\Delta TZ-$$

$$-\underline{TK}+[\underline{G}+\underline{B}-\underline{MD}-\underline{pmg}\cdot\overline{mg}]+$$

$$+[-\overline{ZD}+\overline{C}+\overline{pi}\cdot\bar{i}-\overline{al}-\underline{\overline{TZ}}] \qquad (4.7.19)$$

$$\Delta M=\underline{pmg}\cdot\Delta mg+[\underline{MD}+\underline{pmg}\cdot\overline{mg}]+[-\bar{M}] \qquad (4.7.20)$$

4.8. LAGGED VARIABLES, TRENDS AND SEASONALS

In addition to the 20 current endogenous variables, there are 15 lagged endogenous variables in the model. They are:

$\Delta U_{-2}, \Delta U_{-3}$

$\Delta v_{-2}, \Delta v_{-3}$

$\Delta ZD_{-2}, \Delta ZD_{-3}, \Delta ZD_{-4}, \Delta ZD_{-5}, \Delta ZD_{-6}$

Δpc_{-2}

$\Delta Z_{-2}, \Delta Z_{-3}, \Delta Z_{-4}, \Delta Z_{-5}, \Delta Z_{-6}$

Multiple lags do not fit the model set forth in Chapter 2, but they can be readily converted. This will be shown by means of an example, viz., the transformation of ΔU_{-3}. The single-lagged endogenous variables are numbered from $x(1)$ to $x(20)$ in the order given in Table 4.1, so that

$$\Delta U_{-1}=x(12)_{-1}=x(21) \qquad (4.8.1)$$

where a new variable $x(21)$ has been created. Similarly,

$$\Delta U_{-2}=x(21)_{-1}=x(22) \qquad (4.8.2)$$

In this way, 15 new variables have been created, numbering from $x(21)$ to $x(35)$. As mentioned before (Section 2.8), the system will be controllable if the matrix

$$[C \quad AC \ldots A^{n-1}C]$$

has full row rank. This will be assured if C has full rank, but if 'updating' equations like (4.8.1) and (4.8.2) are used without disturbances, this is no longer so. It is true that Halanay (1969) has shown that in the case of systems of differential-difference equations, controllability is not affected by the appearance of lags, but to be on the safe side we have introduced small perturbations in the updating equations. It was also believed that doing so might speed up the convergence of the S-matrix to a constant $\bar{S}$-matrix, when starting from an arbitrary positive-definite S_0 (cf. Section 2.8 and Chapter 5).

Each of the 10 behaviour equations contains a trend variable; these

10 variables are numbered $x(36)$ through $x(45)$. They are kept unchanged from quarter to quarter by means of equations such as

$$x(36)=x(36)_{-1}+\xi x(36)_{-1} \tag{4.8.3}$$

where the ξx have been added for the reasons explained above for lagged variables.

The 10 behaviour equations also contain seasonal variables which are updated from earlier values. The first equation, for instance, concerning ΔP, contains the seasonal variable $x(47)$. This is related to $x(46)$ and $x(48)$ by the following system

$$\begin{bmatrix} x(46) \\ x(47) \\ x(48) \end{bmatrix} = \begin{bmatrix} 0 & 1 & 0 \\ 0 & 0 & 1 \\ -1 & -1 & -1 \end{bmatrix} \begin{bmatrix} x(46)_{-1} \\ x(47)_{-1} \\ x(48)_{-1} \end{bmatrix} + \begin{bmatrix} 0 \\ 0 \\ 1 \end{bmatrix} \cdot \xi x(48)_{-1} \tag{4.8.4}$$

This system ensures that four seasonal additive factors, the sum of which is zero, will turn up successively. This can be seen by following the values of $x(46)$, $x(47)$ and $x(48)$ through successive quarters:

$$\begin{array}{llll} x(46)=a & b & c & d \\ x(47)=b \rightarrow & c \rightarrow & d \rightarrow & a \\ x(48)=c & d & a & b \end{array}$$

where $d=-(a+b+c)$.

For similar reasons as given above for lagged endogenous variables, disturbance terms were added but it was thought that one disturbance term would be enough to ensure controllability and accelerate convergence to $\bar{S}$.

As there are 10 reaction equations each of which is connected to three seasonal variables there are 30 such variables in all, numbered from $x(46)$ to $x(75)$.

5. Observations and preliminary computations

5.1. INTRODUCTION

The model described in Chapter 4 will be used for experiments ranging over the years 1961–1970. For this purpose, statistical data are required, which are described in the present chapter. These data can rarely be used as such, and the conversions required are explained below.

The data are used for two purposes: to provide 'reference values', around which linearization takes place, and as inputs to a filter-and-control system as described in Chapter 2. For the latter purpose, a linear operator H is specified which maps the state variables into the space of observations. The covariance matrix R of the observation noise entering at this point is also given.

Some preliminary computations that can be executed after the reference values have been introduced are now reported, to unburden later chapters. The model parameters which have now been numerically specified can be used to provide a steady-state covariance matrix $\bar{S}$ through numerical convergence. As a conclusion to the chapter, the dynamic stability properties of the model are briefly analyzed.

5.2. DESCRIPTION OF THE DATA

The Netherlands Central Bureau of Statistics furnishes empirical economic data in a systematic and organized fashion. Nearly all data are classified in a limited number of groups, such as 'H. Labour Market'. We shall use this classification below.

Some series appear monthly, other ones quarterly, with lags that never surpass one quarter. Certain data are provided annually only, and there is one important series (index numbers of wages earned per hour) which is bi-annual and appears with delays of about 6 months.

In the table below only a small subset of the available data will be used.

An attempt to use more would require the same analytical approach, at a higher computational cost, with only slightly more information value.

The table below lists the series used in the present study. More details are provided in Appendix A.

Table 5.1. Statistical time series

Category and description	*Periodicity*	*Source*	*Units*
H. *Labour Market*			
Unemployment			
H1. Men	M	SB	1000 persons
H2. Women	M	SB	1000 persons
I. *Firms and Plants*			
Gross investment in fixed assets (by destination)			
I1. Total private investment	Q	SB	mln. gld.
I2. Total public investment	Q	SB	mln. gld.
Quantity index numbers, by destination			
I3. Total private investment	Q	SB	per cent
I4. Total public investment	Q	SB	per cent
I5. Total private and public investment	Q		
K. *Industry*			
Index numbers of industrial production			
K1. Total for all industries, excluding construction	M	SB	per cent
M. *Foreign Trade*			
Index numbers of foreign trade			
M1. Value of imports	M	SB	per cent
M2. Value of exports	M	SB	per cent
M3. Quantity of imports	M	SB	per cent
M4. Quantity of exports	M	SB	per cent
M5. Price of imports	M	SB	per cent
M6. Price of exports	M	SB	per cent
Imports and exports in mln. gld.			
M7. Total imports	M	SB	mln. gld.
M8. Total exports	M	SB	mln. gld.
P. *National Accounts*			
National Accounts 1970			
Table 9. Sectorwise national income formation			
P1. Indirect taxes	A	NA	mln. gld.
Table 10. National incomes, alternative definitions			
P2. National product (gross; at market prices)	A	NA	mln. gld.
Table 11. National income, primary distribution			
P3. Wages, salaries and social benefits	A	NA	mln. gld.
Table 14. National income and national expenditure			
P4. Private consumption	A	NA	mln. gld.

Table 5.1. *(continued)*

Category and description	*Periodicity*	*Source*	*Units*
P5. Public consumption	A	NA	mln. gld.
P6. Net public investment in fixed assets	A	NA	mln. gld.
Table 15. National means of payments and national expenditure (in constant prices; price base 1963 = 100)			
P7. Private consumption	A	NA	mln. gld.
P8. Public consumption	A	NA	mln. gld.
P9. Net public investment in fixed assets	A	NA	mln. gld.
P10. Increase in stocks and work-in-progress	A	NA	mln. gld.
Table 16. Goods and services totals			
P11. Gross private investment in fixed assets	A	NA	mln. gld.
Table 17. Goods and services totals (in constant prices, price base 1963 = 100)			
P12. Gross private investment in fixed assets	A	NA	mln. gld.
P13. Gross national expenditure	A	NA	mln. gld.
Table 18. Price indices of goods and services totals			
P14. Import of goods and services	A	NA	per cent
Table 47. Main transactions with foreign countries (current transactions)			
P15. Export of goods (f.o.b.)	A	NA	mln. gld.
P16. Import of goods (c.i.f.)	A	NA	mln. gld.
P17. Surplus in current transactions (= income surplus)	A	NA	mln. gld.
Table 51. Quantity and price index numbers of the goods (f.o.b.) and services exports			
P18. Total (public and private) export of goods	A	NA	index number
Table 52. Work force distribution in sectors and firms and registered labour reserve			
P19. Total registered labour reserve	A	NA	1000 man-years
P20. Total work force	A	NA	1000 man-years
Table 53. Wage earners			
P21. Total wage earners in private and public service	A	NA	1000 man-years
Q. *Balance of payments* (transaction basis; current account)			
Q1. Surplus or deficit in current account (cash basis, Netherlands Central Bank)	Q	SB	mln. gld.
R. *Public Finance*			
Taxes on incomes and wealth			
R1. Income tax	M	MF	1000 gld.
R2. Wage tax	M	MF	1000 gld.
R3. Total taxes on incomes and wealth	M	MF	1000 gld.
Cost-increasing taxes			
R4. Total of cost-increasing taxes	M	MF	1000 gld.
Tax yields			
R5. Income tax	A	SRF	mln. gld.
R6. Wage tax	A	SRF	mln. gld.

able 5.1. (continued)

ategory and description	*Periodicity*	*Source*	*Units*
R7. Total taxes on incomes and wealth	A	SRF	mln. gld.
. *Consumption*			
Index numbers of the consumption expenditure of the Netherlands population; total consumption			
T1. Value of consumption	M	SB	index number
T2. Price of consumption	M	SB	index number
T3. Quantity of consumption	M	SB	index number
T4. Quantity per head	M	SB	index number
J. *Price*			
Price indices of the cost-of-living of manual and office workers			
U1. Total (for all workers)	M	SB	per cent
/. *Social Affairs*			
Index numbers of the wages earned per hour by industrial workers			
V1. Total	B	SB	per cent
Percentage increase in wages per workers			
V2. Total (for all workers)	Q	SB	per cent

M: monthly — MF: Maandstatistiek van het Financiewezen (Central Bureau of Statistics)
Q: quarterly — NA: Nationale Rekeningen (National Accounts 1970)
B: bi-annual — SB: Statistisch Bulletin (Central Bureau of Statistics)
A: annual — SRF: Statistiek der Rijksfinanciën (Central Bureau of Statistics)

5.3. CONVERSION OF DATA FOR USE IN THE MACRO-ECONOMIC MODEL

5.3.1. *General*

As usual, the statistical data furnished by the Central Bureau of Statistics requires some conversions, before it can be used with the macro-economic model described in Chapter 4. There are several reasons for this. To begin with, a good deal of the statistical information pertains to monthly periods whereas the model variables refer to economic activity over a quarter of a year. On the other hand, some data are available only on a yearly basis. The units of measurements used by the Bureau are, in general, different from those proposed for the model. In this section, the conversion of statistics furnished by the Bureau into time series used for computations in this study is summarized.

5.3.2. *Data on reference values*

The mathematical model described in Chapter 4 is initially a non-linear one in that identities involving value, volume and prices appear. It was explained that a linearized approximation of the model is obtained by retaining only the first order variations around a given set of 'reference values' of the endogenous variables. The linearized form is also given in Chapter 4. The reference values around which linearization is performed are denoted by a bar above the symbol of the corresponding variable, thus $\bar{\alpha}$ for the variable α. One set of reference values was employed for the four quarters of a calendar year. However, different sets were used for different years because the economic variables display appreciable changes during the years 1961–1970. As the computations were performed over the 40 quarters of this period, 10 sets of reference values, one corresponding to each year were obtained from the statistical data.

The manner of deriving the reference values is indicated in brief now, referring to the labels of the statistical series appearing in Table 5.1. The conversion to appropriate measurement dimensions is omitted because it is simple and need not cause any ambiguity. However, it should be remembered that the units of measurements as described in Chapter 4 were employed throughout the study. Table 5.2 lists the sources of reference values.

5.3.3. *Derivation of observation sequences*

A. *Quarterly Observations*

1. *Unemployment*, $y(1)$

Statistical time series H1 and H2 of Table 5.1 were combined, with a suitable change of units, and monthly averages over quarters were taken to obtain the quarterly series $y(1)$. This is regarded as observations of unemployment U as generated by the macro-economic model of Chapter 4.

2. *Value of gross private investment*, $y(2)$

Series I1 in Table 5.1 refers to observations of quarterly gross private investment in mln. gld. This series was converted to bln. gld. and the quarterly figures were multiplied by 4 to obtain the annual rates. The resulting series $y(2)$ was considered as observations on the values of the gross private investment variable (I) of the macro-economic model.

Table 5.2. Sources of reference values

Variable	*Source*	*Variable*	*Source*
Endogenous		*Exogenous*	
(1) $\bar{P}$	P20	(1) $\overline{az}$	$\bar{P}-\bar{a}-\bar{U}$
(2) $\bar{a}$	P21	(2) $\bar{g}$	P8 + P9
(3) $\bar{C}$	P4	(3) $\bar{G}$	P5 + P6
(4) $\bar{i}$	P12	(4) $\bar{b}$	$P18 \times \bar{B}_{1963}$
(5) $\bar{n}$	P10	(5) $\bar{B}$	P15
(6) $\overline{mg}$	P16/P14	(6) $\overline{MD}$	P15 − P16 − P17
(7) $\bar{l}$	$\bar{L}/\bar{a}$	(7) $\overline{pmg}$	P14
(8) $\overline{pc}$	$\bar{C}/\bar{c}$	*Control*	
(9) $\overline{pi}$	$\bar{I}/\bar{i}$	(1) $\overline{TL}$	$0.1 \times R5 + R6$
(10) $\overline{TZ}$	$R7-\overline{TL}$	(2) $\overline{TK}$	P1
(11) $\bar{v}$	P13		
(12) $\bar{U}$	P19		
(13) $\bar{c}$	P7		
(14) $\bar{I}$	P11		
(15) $\bar{L}$	P3		
(16) $\bar{Z}$	$\bar{Y}-\bar{L}-\overline{TK}$		
(17) $\bar{Y}$	P2		
(18) $\overline{LD}$	$\bar{L}-\overline{TL}$		
(19) $\overline{ZD}$	$\bar{Z}-\overline{TZ}$		
(20) $\bar{M}$	$P16+\overline{MD}$		

3. *Volume of gross private investment, $y(3)$*

The series I3, which refers to the indices of the volume of gross private investment in each quarter, is used to derive observations on the corresponding variable (i) of the model. Since the indices are not given for a common base for the period 1961–1970, the statistical data were converted by a simple ratio transformation to a common base corresponding to the year 1963. With the help of the 1963 volume of investments in bln. gld. the indices were then transformed to quarterly observations $y(3)$.

4. *Volume of gross national expenditure, $y(4)$*

The monthly index numbers of industrial production, designated by K1 in Table 5.1, were employed to yield observations on the variable v in the model. The monthly indices were averaged to obtain quarterly figures;

which were then converted to a common base corresponding to an index of 100 for the year 1963. From 1963 to 1969, the annual gross national expenditure at constant prices changed by 41 percent approximately while the index in series K1 rose by 59 percent. The ratio of the rise in K1 to that in the gross national expenditure is then 41/59 or roughly 0.7. Therefore, to obtain $y(4)$, the quarterly production indices were multiplied by 0.7 times the 1963 volume of gross national expenditure in bln. gld. and 0.3 of the same were added as a constant.

5. *Volume of goods imports,* $y(5)$
The statistical indices of quantity of goods imports labeled M3 in Table 5.1 were adopted to give quarterly observations $y(5)$ on the volume of goods (*mg*) in bln. gld. The monthly series M3 were first averaged to result in quarterly figures which were transformed to a common base of 100 for 1963 and then to figures in bln. gld.

6. *Value of private consumption,* $y(6)$
Series T1 of Table 5.1, comprising monthly indices of private consumption, were considered as observations on the variable C. The average over a quarter of the monthly indices was obtained. The resulting quarterly indices were then converted by a simple ratio transformation such that the year 1963 had an index of 100. The value of private consumption in 1963 appearing in the national account was employed to convert the indices into figures with bln. gld. units.

7. *Price of private consumption,* $y(7)$
The monthly cost-of-living price indices series U1 were regarded as information on the cost of consumption goods (*pci*). Quarterly indices were obtained by taking averages of the monthly figures. The resulting indices were, in turn, converted such that the final prices had a base value of 1 for the year 1963.

8. *Wage rate,* $y(8)$
The quarterly series V2 in Table 5.1 gives percentage rise in wages compared to those in the month of October. The series was converted to represent wage rises in bln. gld. per 100,000 wage earners by multiplying with appropriate wage levels. The outcome was regarded as observations of the variable l.

9. *Value of imports of goods,* $y(9)$
The monthly series M7 was taken to yield observations of the variable M,

which represents total imports, i.e., including net imports of services. Since the services imports in excess of exports are exogenous, the trade figures referring to imports of goods are appropriate to give information on the total imports. The series M7 was transformed to quarterly figures in bln. gld. at annual rates.

10. *Taxes on non-wage income*, $y(10)$
To obtain an observation series on the variable TZ, the tax on wage income series R2 was subtracted from the series R3, which represents total taxes on incomes and wealth. Of the income tax series R1, 10 per cent was transferred to the control variable $\overline{TL}$. The quarterly observations were expressed in bln. gld. at annual rates.

B. *Bi-annual Observations*

11. *Wage rate*, $y(11)$
Series V1 in Table 5.1 has a bi-annual periodicity. It consists of wage rates per hour compared to the wages in the preceding month of October. This series was also regarded as information on the variable l. For this purpose V1 was transformed to figures in bln. gld./100,000 persons with the help of the corresponding reference values $\bar{l}$.

C. *Annual Observations*

Eight of the reference values, which, as indicated earlier, were derived from the national accounts except for $\overline{TZ}$, were considered as annual observations on some of the variables of the macro-economic model. These annual observations represent totals of the quarterly flow variables over a year. The labels of these observations and the variables on which they give information are indicated below.

12. *Value of consumption*, $y(12)=\bar{C}$

13. *Volume of gross private investment*, $y(13)=\bar{i}$

14. *Increase of stocks*, $y(14)=\bar{n}$

15. *Volume of gross national expenditure*, $y(15)=\bar{v}$

16. *Volume of consumption*, $y(16)=\bar{c}$

17. *Value of gross private investment*, $y(17)=\bar{I}$

18. *Wage income*, $y(18)=\bar{L}$

19. *Non-wage income*, $y(19)=\bar{Z}$

5.4. OBSERVING THE SYSTEM STATE

In the mathematical model of the Netherlands economy described in Chapter 4, the economic state is characterized by a set of 20 endogenous variables. As mentioned in the previous sections, the statistical information provided by the Central Bureau of Statistics does not readily correspond to the state variables in the model. The units of measurements and periodicity do not conform with those selected for this study. However, it is of more significance that observation data are not available each quarter for all the variables involved in the mathematical model. As noted in Section 5.3 only 10 time-series are selected as quarterly observations on the system state. There is one observation series which is bi-annual in periodicity, and eight refer to annual totals only.

Quarterly Observations

In Section 5.3, the manner of obtaining observations $y(1)$ through $y(10)$ is described. These series become available with delays not exceeding three months in general. They can then be employed to update estimates of the system state. The relationship of the quarterly observations with the system variables is now considered. It is assumed that, in addition to state variables, an error of observation enters into the statistical information.

$$y(1)=x(12)+\eta(1)+\bar{U} \tag{5.4.1}$$

$$y(2)=x(14)+\eta(2)+\bar{I} \tag{5.4.2}$$

$$y(3)=x(4)\ \ +\eta(3)+\bar{i} \tag{5.4.3}$$

$$y(4)=x(11)+\eta(4)+\bar{v} \tag{5.4.4}$$

$$y(5)=x(6)\ \ +\eta(5)+\overline{mg} \tag{5.4.5}$$

$$y(6)=x(3)\ \ +\eta(6)+\bar{C} \tag{5.4.6}$$

$$y(7)=x(8)\ \ +\eta(7)+\overline{pc} \tag{5.4.7}$$

$$y(8)=x(7) \quad +\eta(8)+\bar{l}-\bar{l}_{-1} \tag{5.4:8}$$

$$y(9)=x(20)+\eta(9)+\overline{M}-\overline{MD} \tag{5.4.9}$$

$$y(10)=x(10)+\eta(10)+\overline{TZ} \tag{5.4.10}$$

Bi-annual Observations

There is only one bi-annual observation sequence, $y(11)$, in the set of statistical data considered in the study. This relates to variable l, the wage rate. The deviation of l from its reference value $\bar{l}$ is denoted by $x(7)$. Given that $y(11)$ refers to the first two quarters of a year, the following relationship is assumed for its generation.

$$y(11)=0.5(x(7)+x(7)_{+1})+\eta(11)+\bar{l} \tag{5.4.11}$$

where $x(7)$ and $x(7)_{+1}$ represent deviations Δl over the first two quarters of a year. With transition of time, a similar relationship holds for the bi-annual $y(11)$ which refers to the latter two quarters of the year.

Annual Observations

The annual observations $y(12),\ldots, y(19)$ relate to totals of some of the quarterly flow variables of the macro-economic model. They are, therefore, available only once a year. Their relationship with the model variables is as follows (the symbols $\eta(12)$–$\eta(19)$ represent errors of observation):

$$y(12)=0.25(x(3)+x(3)_{+1}+x(3)_{+2}+x(3)_{+3})+\eta(12)+\bar{C} \tag{5.4.12}$$

$$y(13)=0.25(x(4)+x(4)_{+1}+x(4)_{+2}+x(4)_{+3})+\eta(13)+\bar{\imath} \tag{5.4.13}$$

$$y(14)=0.25(x(5)+x(5)_{+1}+x(5)_{+2}+x(5)_{+3})+\eta(14)+\bar{n} \tag{5.4.14}$$

$$y(15)=0.25(x(11)+x(11)_{+1}+x(11)_{+2}+x(11)_{+3})+\eta(15)+\bar{v} \tag{5.4.15}$$

$$y(16)=0.25(x(13)+x(13)_{+1}+x(13)_{+2}+x(13)_{+3})+\eta(16)+\bar{c} \tag{5.4.16}$$

$$y(17)=0.25(x(14)+x(14)_{+1}+x(14)_{+2}+x(14)_{+3})+\eta(17)+\bar{I} \tag{5.4.17}$$

$$y(18)=0.25(x(15)+x(15)_{+1}+x(15)_{+2}+x(15)_{+3})+\eta(18)+\bar{L} \tag{5.4.18}$$

$$y(19)=0.25(x(16)+x(16)_{+1}+x(16)_{+2}+x(16)_{+3})+\eta(19)+\bar{Z} \tag{5.4.19}$$

In these equations, the state variables, like $x(3),\ldots, x(3)_{+3}$, refer to the four quarters of a year for which the observations are available.

The random disturbances appearing in Equations (5.4.1)–(5.4.19) were, in conformity with the assumptions in Chapter 2, assumed to have zero mean and to be uncorrelated over time. The correlations over the same time period among various observations were ignored. Correlation between quarterly and annual observation disturbances will occur when the same basic statistics are used. For the sake of simplicity, these correlations were also neglected. In short, each of the disturbances $\eta(1)$–$\eta(19)$ were assumed to be zero mean, uncorrelated processes with fixed variances. Their variances, given in Table 5.3, were chosen on aprioristic considerations.

Table 5.3. Variance of observation errors

No.	*Random Disturbance*	*Variance*
1	$\eta(1)$	0.01
2	$\eta(2)$	4.0
3	$\eta(3)$	4.8
4	$\eta(4)$	22.0
5	$\eta(5)$	0.36
6	$\eta(6)$	16.0
7	$\eta(7)$	0.0004
8	$\eta(8)$	0.0008
9	$\eta(9)$	0.36
10	$\eta(10)$	0.0001
11	$\eta(11)$	0.0004
12	$\eta(12)$	4.0
13	$\eta(13)$	1.04
14	$\eta(14)$	0.09
15	$\eta(15)$	5.50
16	$\eta(16)$	4.25
17	$\eta(17)$	1.00
18	$\eta(18)$	2.00
19	$\eta(19)$	6.00

5.5. DETERMINATION OF THE COVARIANCE MATRIX S

As all the necessary numerical assumptions have now been made, the conditional covariance matrices S_i can now be obtained. This task would be simpler if the model were stationary over the entire period envisaged (1961–1970) because then the Kalman-Bucy theorem on stability of the filter could be applied. However, within each calendar year the reference values do not change, and therefore the A_i matrices are constant within the years as well, so that the model can be split in 10 sub-models, one for each year, which are stationary.[1]

Within each year, the Kalman-Bucy theorem applies. Of its two main conditions, controllability and observability, the first one would be assured if it could be shown that the matrix

$$[C \quad AC \dots A^{n-1}C]$$

had full row rank. As several rows of C are empty, full row rank cannot be assumed to exist by referring to C only. However, it is reasonable to expect that the Halanay (1969) result which was given in Section 4.8 applies in this case, so that controllability can be assumed to obtain.

As regards observability, the criterion that the matrix

$$[H^T (HA)^T \dots (HA^{n-1})^T]$$

has full row rank has not been shown to be satisfied, because the considerable amount of computation involved did not seem worthwhile. However, the least-squares algorithm used to obtain the initial state vector x_0 (cf. Section 7.6) requires full row rank of a matrix very similar to the above to be successful, and as this computation succeeded, full row rank and thereby observability can reasonably be assumed to obtain.

However this may be, for 1963 numerical convergence to a constant S-matrix starting from an identity matrix was obtained, using the recursion

$$P = ASA^T + CQC^T$$

$$K = PH^T (HPH^T + R)^{-1}$$

$$S = (I - KH)\, P$$

It cannot be stated for certain that this matrix is unique, because the conditions of the Kalman-Bucy theorem have not been shown to be fulfilled, but the S-matrix obtained is good enough for the purpose of our experiments.

For the other years, similar converging computations could have been made. Alternatively, a constant S-matrix for 1961 could have been obtained from which the S-matrices for subsequent years could have been derived by recursion.

Neither of these two possibilities has been explored, and in fact the S-matrix obtained for 1963 was used for all years without modification.

1. The prices of goods imports, pmg, were given the corresponding annual reference values $\underline{\overline{pmg}}$.

As the Q and R matrices are identical for all years, it may be hoped that S-matrices derived for individual years will not differ much from each other. In addition, it may perhaps be said that choosing a particular S-matrix on aprioristic grounds is no more arbitrary than choosing a Q or R matrix in the same manner.

The minimum-variance property of the Kalman filter remains, whatever S-matrices are used to obtain the error-correction matrices K. Still, some S-matrices will lead to smaller expected errors than other S-matrices, but it is likely that the effect of using not too different S-matrices on the expected variance is small.

A minor sophistication should, however, be mentioned. The observation data described in this chapter contain annual and bi-annual time-series as well as quarterly ones. For short-run control policy determination, as discussed in Chapter 7, the annual and the bi-annual observations arrive too late, and the policy has to be based, in the formulation of this study, entirely on quarterly observations. With this complication in mind, two steady-state $\bar{S}$ matrices were obtained. In one, all the observations were included. In the other, only quarterly data were taken into account. In the Kalman filter computations in Chapter 7, the updating matrix K_i was obtained from the latter $\bar{S}$ matrix.

5.6. THE PREDICTION-CORRECTION MATRIX K

If the matrix $\bar{S}$ is obtained by numerical convergence in the manner described in the preceding section, the prediction-correction matrix K is obtained at the same time as it forms one link of the recursion used. This matrix converts the differences between observations and predictions into corrections of the predictions of the state vector:

$$x_i^* = \hat{x}_i + K[y_i - H\hat{x}_i]$$

The more interesting part of it is shown in Table 5.4. The rows not shown refer to lagged variables, trends and seasonals.

The first column of the table shows that the observed unemployment $y(1)$ strongly affects the estimated unemployment U: every 1000 unemployed observed above the predicted number lead to a correction of the estimated unemployment by 982. This is due to the very small observation noise assumed in the unemployment statistics. It is interesting to see that the model allocates 592 of the additional unemployment to an upward correction in the estimated labour force P, and the remaining 389 to a reduction in employment a.

Table 5.4. Prediction-correction matrix K

	$y(1)$	$y(2)$	$y(3)$	$y(4)$	$y(5)$	$y(6)$	$y(7)$	$y(8)$	$y(9)$	$y(10)$
P	0.592	−0.016	−0.066	0.036	0.041	0.071	1.268	−2.992	0.040	−0.800
a	−0.389	−0.016	−0.066	0.036	0.041	0.071	1.261	−2.992	0.041	−0.802
C	−0.280	0.041	−0.046	0.110	0.071	0.219	8.757	13.885	0.070	−0.004
i	0.010	0.137	0.264	0.042	0.018	−0.013	1.346	−1.359	0.018	0.368
n	−0.131	0.004	−0.026	0.090	−0.115	−0.013	0.557	−0.076	−0.114	0.030
mg	−0.007	0.001	0.001	−0.000	0.480	0.001	−0.116	0.037	0.473	0.009
l	−0.000	0.000	−0.000	0.000	0.000	0.000	0.197	0.758	0.000	−0.001
pc	0.000	0.000	0.000	−0.000	−0.000	0.000	0.813	0.098	−0.000	0.000
pi	−0.001	0.027	−0.013	−0.001	−0.000	0.001	−0.002	0.171	−0.000	0.001
TZ	0.000	0.000	0.000	0.000	0.000	−0.000	0.000	−0.000	0.000	0.999
v	−0.366	0.172	0.194	0.238	−0.032	0.152	−16.432	7.249	−0.031	0.374
U	0.982	−0.000	0.000	−0.000	−0.000	−0.000	0.006	−0.000	−0.000	0.001
c	−0.246	0.030	−0.043	0.103	0.065	0.180	−18.373	8.703	0.064	−0.022
I	−0.015	0.540	0.114	0.031	0.011	0.010	1.516	0.777	0.011	0.450
L	−0.427	−0.012	−0.081	0.050	0.048	0.104	8.736	24.952	0.047	−0.934
Z	0.008	0.597	0.121	0.182	−0.581	0.110	2.218	−10.402	−0.572	1.402
Y	−0.418	0.584	0.040	0.233	−0.532	0.214	10.955	14.549	−0.524	0.467
LD	−0.427	−0.012	−0.081	0.050	0.048	0.104	8.736	24.952	0.047	−0.934
ZD	0.008	0.597	0.121	0.182	−0.581	0.110	2.218	−10.402	−0.572	0.402
M	−0.007	0.001	0.001	−0.000	0.473	0.001	−0.114	0.036	0.519	0.009

Observations of the consumer price index $y(7)$ affect the estimate o. almost proportionately (the relevant element is 0.813). This is distributed over value C and volume c of private consumption such that for every 1% of correction in the consumer price index, the volume estimate is reduced by 184 million (the coefficient shown is -18.373) while the value estimate increases by 88 million.

If profit taxes $y(10)$ bring in 100 million more than expected, this leads to an upward correction in profits Z of no less than 140 million. The national product Y is also increased, by 47 million, but wages L bear the difference of 93 million. Very little of this is attributed to lower wage rates l and almost all of it is reflected in a decline of estimated employment by 8020 persons, which, however, hardly affects the unemployment estimate, as the labour force estimate a is reduced by 8000 persons.

5.7. DYNAMIC STABILITY

Once the A-matrix has been numerically defined it becomes possible to ascertain whether the linear system is dynamically stable in the sense that as time goes to infinity there is no danger that the system will move outside a finite range.[2] This kind of stability is considered essential if the model is to give a realistic representation of an existing economic system. It is different from the stability concept used in filter theory (cf. Section 2.8).

Table 5.5. Dominant characteristic roots of the macro-economic model, 1963

No.	*Real Part* α	*Imaginary Part* ω	*Damping Factor* $\sqrt{\alpha^2+\omega^2}$	*Time Period in Quarters* $T=2\pi/\tan^{-1}(\omega/\alpha)$
1	0.983	0.0	0.983	–
2	0.740	±0.317	0.805	15.53
3	−0.738	0.0	0.738	–
4	−0.253	±0.668	0.715	5.20
5	0.361	±0.609	0.708	6.06
6	−0.356	±0.494	0.609	6.64
7	0.552	0.0	0.552	–
8	−0.421	±0.274	0.503	10.89
9	0.080	±0.536	0.542	4.42
10	0.275	0.0	0.275	–

2. For a more exact definition of stability of a system in the Lyapunov sense see, e.g., Freeman (1965), page 157.

Dynamic stability is assured if and only if all the roots of the A-matrix lie within the unit circle. Table 5.5 lists the largest roots for the A-matrix of 1963, showing 4 real roots and 6 pairs of conjugate complex roots.

The largest root of the system is found to be real with magnitude less than one. This implies that the system is stable. The damping is, however, not very strong.

The system has a number of periodic modes of behaviour. The largest time period is as big as 15 quarters with a damping factor of about 0.8. This introduces considerable cyclical components in the response of the model. The other periodic components with not too small damping factors have time periods of about 11, 7, 6, 5 and 4 quarters. That is, the model has a variety of oscillatory modes.

6. Projection of exogenous variables

6.1. INTRODUCTION

If a state transition model is to be employed to predict macro-economic activity in the Netherlands, followed by control action, the future values of the exogenous variables, as their name signifies, need be supplied from outside the model. In real life, these predictions will often be derived from an analysis of the systems that generate these exogenous variables, and from special information: but for the experiments reported here, no such sources are available. The prediction of an exogenous variable is affected only by present and past observations of the same variable. The Kalman filter provides a convenient tool for such predictions, and given suitable assumptions it assures minimum error variance. Its general principles have been set out in Chapter 2; the following sections will show how these are applied for the purpose of projection of exogenous variables.

The prediction process requires an estimate of the initial state vector. This was obtained from the observations by another numerical convergence process described below. The same process provided 'smoothed' estimates, which were used as exogenous inputs in the experiments of Chapter 7.

If quarterly observations are not available, annual observations must be used to provide estimates of quarterly values. This is explained in the final section.

6.2. PROJECTION OF QUARTERLY TIME-SERIES, ASSUMING SEASONAL VARIATION

In the macro-economic model described in Chapter 4, the quarterly value of exports $\underline{B}$ is one of the exogenous variables. Trade statistics are available monthly, and quarterly figures can be derived from them. Series M8 in Table 5.1 is employed for this purpose. Exports, both on a monthly as well as on a quarterly basis, exhibit seasonal variations. Therefore, in a technique for projecting exports into the future, it is appropriate to include a seasonal pattern. The discussion below applies equally to the quarterly volume of exports $\bar{b}$.

A discrete-time state-transition model can be devised to describe the generation of a time-series with trend and seasonal variation. The model described here is on the same lines as the seasonal variation model employed for the endogenous variables, explained in Section 4.8. The essential relationships are

$$x(k)=x(k)_{-1}+dx(k)_{-1}+sx(k) \tag{6.2.1}$$

$$y(k)=x(k)+\eta(k) \tag{6.2.2}$$

where $x(k)$ is the level of the time-series at time t, and $dx(k)$ and $sx(k)$ its linear trend and seasonal components, respectively. In conformity with the model in Chapter 2, the observations $y(k)$, which form the time-series, contain random disturbances $\eta(k)$. The trend and seasonal variables are assumed to have transitions similar to those described in Section 4.8. The complete system can be written in the following vector-matrix form:

$$\begin{bmatrix} z(1) \\ z(2) \\ z(3) \\ dx(k) \\ x(k) \end{bmatrix} = \begin{bmatrix} 0 & 1 & 0 & 0 & 0 \\ 0 & 0 & 1 & 0 & 0 \\ -1 & -1 & -1 & 0 & 0 \\ 0 & 0 & 0 & 1 & 0 \\ 0 & 1 & 0 & 1 & 1 \end{bmatrix} \begin{bmatrix} z(1)_{-1} \\ z(2)_{-1} \\ z(3)_{-1} \\ dx(k)_{-1} \\ x(k)_{-1} \end{bmatrix} + \begin{bmatrix} 0 & 0 \\ 0 & 0 \\ 1 & 0 \\ 0 & 1 \\ 0 & 0 \end{bmatrix} \begin{bmatrix} \xi sx(k)_{-1} \\ \xi dx(k)_{-1} \end{bmatrix} \tag{6.2.3}$$

where $z(1)=sx(k)$; $z(2)=sx(k)_{+1}$; $z(3)=sx(k)_{+2}$

$$y(k)=[0 \quad 0 \quad 0 \quad 0 \quad 1]\,[z(1), z(2), z(3), dx(k), x(k)]^T+\eta(k) \tag{6.2.4}$$

In Equations (6.2.3) and (6.2.4), the matrices A, C and H are easily recognized.

The random disturbance processes $\xi sx(k)$, $\xi dx(k)$ and $\eta(k)$ are assumed to be uncorrelated zero-mean processes with fixed finite variances. No correlation over time is supposed to exist. This is in accordance with the formulation in Chapter 2. The standard deviations of disturbances in the seasonal and trend transitions were set at nearly 2% of the reference value in 1963 (18 bln. gld.) and that of the errors in observation at 1%.

For this formulation, full row rank for the block matrices $[C \quad AC\ldots \ldots A^4C]$ and $[H^T(HA)^T\ldots(HA^4)^T]$ can be shown to exist. That is, the conditions for controllability and observability are satisfied. Therefore, a unique finite conditional covariance matrix $\bar{S}$ exists. Since all the quantities: A, B, C, H, Q and R are known, the matrix $\bar{S}$ can be determined by numerical convergence starting from a given positive-definite matrix, say, the unit matrix. An analytical approach does not appear advisable. Even in the case when seasonal components are absent (cf. Section 6.3), an analytical solution for the corresponding $\bar{S}$-matrix of order 2 is far from simple (see Vishwakarma (1970a), Theil and Wage (1964)). The numerical procedure gives the following steady-state matrix.

Table 6.1. Steady-state covariance matrix $\bar{S}$ for filtering of quarterly value of exports

	1	*2*	*3*	*4*	*5*
1	0.128	−0.100	−0.014	−0.086	0.016
2	−0.100	0.337	−0.188	−0.087	−0.007
3	−0.014	−0.188	0.375	0.014	−0.003
4	−0.086	−0.087	0.014	0.245	0.013
5	0.016	−0.007	−0.003	0.013	0.031

As the steady-state covariance matrix $\bar{S}$ is known, both filtering and prediction of the quarterly exports can proceed recursively as observations become available, provided that an initial state vector is known. An approach to specify a starting vector is considered in the next section.

6.3. SMOOTHING AND INITIAL STATE VECTOR

In the recursive filtering approach the estimate of the system state is updated each time a fresh observation is available. At the beginning, an initial state vector is required. In the model formulated in Section 6.2, this means that estimates of the trend and seasonals must be known. They can be set on aprioristic grounds, in line with the approach in this study. A numerical method for the purpose of specifying the initial vector is indicated here that takes into consideration a given sample of observations.

It has been noted before that a steady-state covariance matrix $\bar{S}$ exists for the model in Equations (6.2.3)–(6.2.4) given that the random disturbance variances Q and R are constant. This is fulfilled by specification. Hence, the prediction-correction vector K will also be constant. The smoothing matrix J in the Meditch algorithm (Section 2.7), which enables updating of past estimates of the state vector in the light of later observations, then also becomes constant. In order to obtain the initial vector, a first assumption of zero trend and seasonal components and level equal to the observed value was made, and prediction-correction was applied using the Kalman filter over the sample of observations. Starting from the end of the sample period, the Meditch smoothing algorithm was employed to obtain the next guess of the starting vector. The procedure was repeated till numerical convergence was obtained.

Once an initial vector is obtained which remains reasonably unaltered by successive filtering and smoothing over the sample, further filtering leaves all the state vector estimates pertaining to each quarter of the sample almost the same as in the previous run. The level component of the state vector resulting from such an operation was stored for use in further numerical computations with the macro-economic model to be described in Chapter 7. Such a series was termed 'smoothed', in contrast to the observed values of the exogenous variables.

In Figure 6.1, one quarter ahead predictions of the value of exports, based on smoothed state vectors are plotted as well as the corresponding realized values. The prediction errors have a standard deviation of about 5% of the reference level of the value of exports in 1965. This was to be expected inasmuch as the seasonal and trend components are disturbed each quarter by a random term of standard deviation of 2% each and the observation errors have a standard deviation of 1% of the 1965 reference level of the value of exports.

Sections 6.2 and 6.3 relate to a state transition model which generates a seasonal pattern for a quarterly observed time-series. The model can, of course, be simplified to describe a quarterly time series with only a trend and no seasonal by dropping the seasonal components of the state vector. The prediction, filtering and smoothing are, however, analogous to the seasonal case. This was applied in the case of the price of imports.

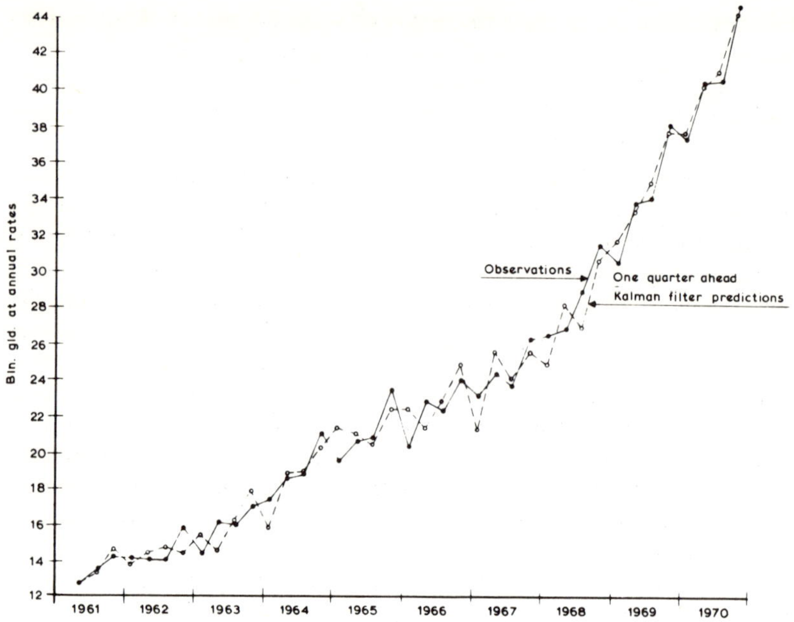

Figure 6.1. Prediction of value of quarterly exports

6.4. ESTIMATION OF QUARTERLY PROCESSES BASED ON ANNUAL TOTALS

In the case of net imports of services (<u>MD</u>) and Government expenditure at current and constant prices (<u>G</u>, <u>g</u>), only the totals over a whole calendar year are obtainable from national accounts. As of now, complete national accounts for each quarter are not published. Therefore, inferences about the quarterly values need be based on annual totals. A hypothetical seasonal movement, that over a year averages to an insignificant value, is necessarily lost in the annual totals. In addition, the sampling theorem of information theory (cf. Blackman and Tukey (1956), Raisbeek (1964)) would imply that the fastest cyclical component that could theoretically be reconstructed from annual observations will have a time-period of at least two years. In practice, the cyclical component with the smallest time-period that could reliably be detected in annual observations will probably have a time-period of at least 8 years. In the short-run, it is therefore reasonable to suppose merely a trend model. In addition, we introduce another simplification by assuming that the state transition and observations have insignificant random disturbances, i.e., the whole noise sequences ξ_i and η_i are supposed to be absent. In mathematical terms, the model appears as follows:

$$\begin{bmatrix} dx(k) \\ x(k) \end{bmatrix} = \begin{bmatrix} 1 & 0 \\ 1 & 1 \end{bmatrix} \begin{bmatrix} dx(k)_{-1} \\ x(k)_{-1} \end{bmatrix} \tag{6.4.1}$$

and

$$y(k) = 0.25(x(k) + x(k)_1 + x(k)_2 + x(k)_3) \tag{6.4.2}$$

In Equation (6.4.2), $y(k)$ stands for the annual total and $x(k)$ has subscripts referring to 4 quarters of the year. The coefficient 0.25 in the observation equation appears because the variables are measured at annual rates.

The trend is easily estimated by the use of ordinary least squares. Since the Kalman filter is a generalized linear minimum-variance unbiased estimator, the least squares estimates can also be obtained in a recursive manner by means of the filter. The result of fitting the model in Equations (6.4.1) and (6.4.2) by the least squares procedure for the exogenous variable $\underline{G}$ is shown in Figure 6.2.

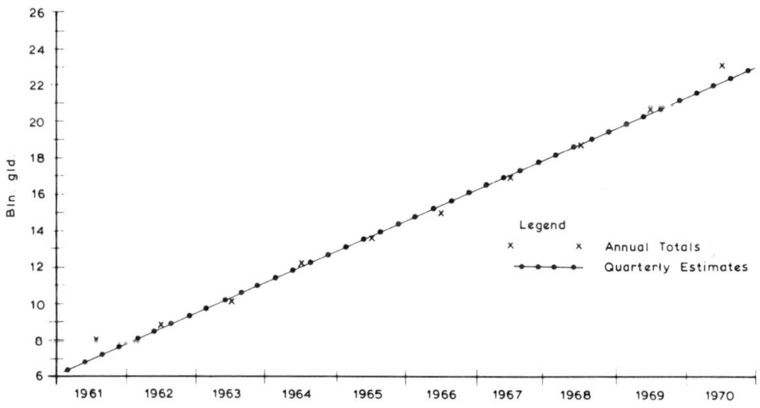

Figure 6.2. Trend estimates of $\underline{G}$ estimated by Ordinary Least Squares

There are other techniques, notably that reported by Boot, Feibes and Lisman (1967), which also attempt to derive quarterly values from annual totals. These are less suitable for prediction, however.

7. Closed-loop control policy experiments

7.1. INTRODUCTION

The materials provided in Chapters 4, 5 and 6 are almost sufficient to set up a closed-loop control system for regulation of the economic activity. Missing elements are the costs of using policy instruments and the objective function. These are discussed below.

The complete system is then employed to run a number of experiments with closed-loop control in the Netherlands in the 1960's.

7.2. A CHOICE OF ECONOMIC POLICY OBJECTIVES

Theil (1964) quotes five goals for economic policy mentioned by the Social and Economic Council (an advisory body) in 1956. These may still be taken as the more important objectives.

Balance-of-payments surplus, prices of consumption goods and unemployment are selected here for regulation. On the other hand, investment, which has implications for production in the future, and distribution of income among different social groups are not included. In a more realistic discussion, such variables will of course arise, and can be handled within the same framework of analysis.

Balance-of-payments

As the Netherlands are a highly open economy, with exports and imports each about 50% of the gross national product, the need for some kind of balance-of-payments equilibrium in transactions with other economies is overwhelming.

The economic objective adopted here is to attain a balance-of-payments surplus in current transactions equivalent to 1% of the gross national product. This surplus serves to finance a possible structural deficit in the capital account, and development aid.

Since export of goods ($\underline{b}, \underline{B}$) and the imports of services net of their exports are exogenous variables in the macro-economic model described in Chapter 4, the balance-of-payments can be adjusted by regulating imports only. For the sake of simplicity, the reference value of the gross national product ($\bar{Y}$) was taken for evaluating the surplus. The desired value of imports in a given quarter can then be expressed as

$$\tilde{M} = \underline{B} - 0.01\ \bar{Y} \tag{7.2.1}$$

where the tilde ($\sim$) distinguishes the desired value from the corresponding endogenous value of the imports. It is to be noted that a seasonal effect is left in exports $\underline{B}$ and, hence, in $\tilde{M}$.

Prices of consumption goods

Frequent and large movements in prices lead to inefficient functioning of the economy. Persons receiving fixed money incomes find their real incomes varying with price fluctuations. Uncertainty in price development gives rise to speculatory reactions which may result in variations in the economic activity.

For the purpose of this study, a smoothly rising price of consumption goods was adopted as an objective. In Figure 7.1, the reference values, $\overline{pc}$, of the consumption price levels are plotted for the years 1961–1970. The straight trend line, drawn free-hand, is taken to represent the desired or 'ideal' time-path for price developments over the period 1961–1970. The desired prices $\tilde{pc}$ can be obtained from the following relation:

$$\tilde{pc} = 0.90 + 0.012 \cdot \{(\text{YEAR} - 1961) \cdot 4 + \text{QUARTER}\} \tag{7.2.2}$$

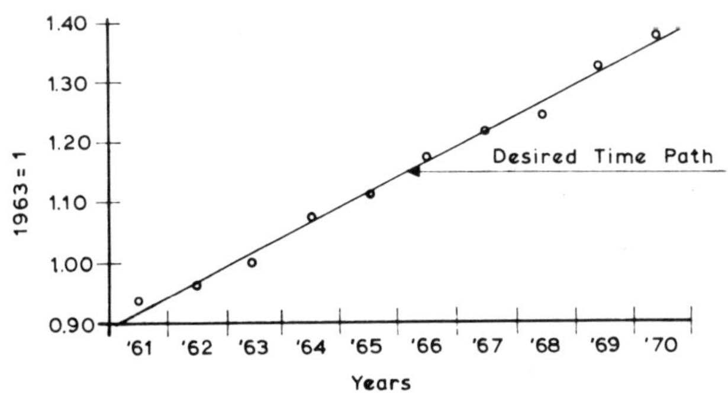

Figure 7.1. Observations and desired time path of consumption prices

where 'YEAR' is any calendar year between 1961 and 1970, and 'QUARTER' takes values 1, 2, 3, or 4 depending on the given quarter of a year.

The desired prices $\tilde{pc}$ generated by Equation (7.2.2) are independent of the seasonal variations. In contrast, the prices pc generated by the macroeconomic model contain seasonal effects. The seasonal variation can be easily introduced since it is also concurrently predicted and updated along with other endogenous economic variables in the linear model. In the scheme of computations reported here, the seasonal effect was added to $\tilde{pc}$ in Equation (7.2.2) so as to obtain a seasonally 'adjusted' desired price level; 'adjusted' here is taken to mean exhibiting seasonal variation. Instead, the variation could, of course, be subtracted from the variable pc generated by the model of Chapter 4 in order to obtain seasonally 'adjusted' prices, in the sense usually understood.

Unemployment rate

A high utilization of labour, which is a factor of production, is desirable in order to optimize production, apart from the social value accorded to employment. When specifying a desirable level of employment, the matter of frictional unemployment need to be considered.

In this study, regulation of the unemployment rate to 1% of the available working population was adopted. The working population was set *a priori* by free-hand smoothing of the curve resulting from plotting the reference values of the variable P, i.e., $\bar{P}$, over 1961–1970, shown in

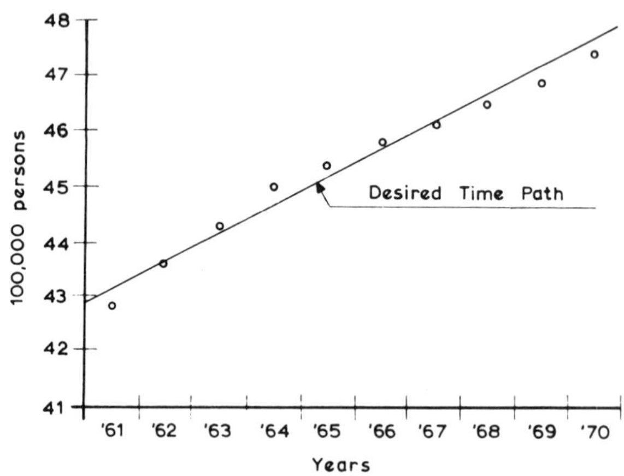

Figure 7.2. Observations and desired time path of total work force

Figure 7.2. The desired level of unemployment, $\tilde{U}$, in any quarter can be expressed by the relation

$$\tilde{U} = 0.01\,\{43.0 + 0.12(\text{YEAR}-1961)\,4 + \text{QUARTER}\} \qquad (7.2.3)$$

where 'YEAR' and 'QUARTER' have the same meaning as for Equation (7.2.2).

Again, the unemployment levels $\tilde{U}$ are free of seasonal variation. In the Netherlands, there is considerable seasonal fluctuation in unemployment. As in the case of prices, the seasonal variation can be appropriately accounted for since it is simultaneously generated in the macro-economic model.

7.3. INSTRUMENTS OF ECONOMIC POLICY

For the short-run regulation of economic activity, such means of policy are suitable that can be adjusted with relative ease and in good time. In the Netherlands, statutory provisions exist – the so-called 'wobble (wiebel)-tax' law – which permit changes in several taxes, within specified limits, by the administration without prior approval of the Parliament. For demonstrating the application of control theory to the determination of short-run economic policy, autonomous changes in the wage, non-wage and indirect taxes, viz., $\underline{TRL}$, $\underline{TRZ}$, and $\underline{TRK}$, were adopted.

Other instruments would have increased the size of the problem and would have been cumbersome from a practical point of view. The government expenditure ($\underline{G}, \underline{g}$), although it has appreciable effect on the overall activity, is difficult to change, particularly downwards in the time spans under consideration. Wage rates have for long time required government approval in the Netherlands, and could be regarded as an instrument of policy in a sense. However, negotiations surrounding wage settlements would make them inconvenient for short-run regulation of economic activity. The effects of monetary instruments, like the interest rate or the money supply, appear to be rather uncertain. Moreover, for convenience, several taxes that exist in practice are grouped under the three categories.

Boundaries to the extent to which the instruments of economic policy may be used can arise out of psychological, political or other considerations (cf. Tinbergen (1956), p. 59), and even be stipulated by law, as in the case of the 'wobble tax'. Such bounds were used, *inter alia*, by Van Eijk and Sandee (1959). In the present study, quadratic penalties on the use of instruments are introduced instead (cf. Theil (1964)) because certainty equivalence is then preserved.

The model predicts the response of the economy to the use of these instruments. Two different types of instrument use are distinguished in literature: the 'unit impulse', when an instrument is applied to the extent of one unit for one time-period only, and then reverts to its original value, and the 'unit step', when an instrument is applied to the extent of one unit, and stays at that value. In Figure 7.3, the responses to a unit step starting in quarter 1 of 1963 in each of the three instruments are plotted.

These responses are shown for the three endogenous variables entering the objective, viz., imports M, consumer prices pc, and unemployment U. They are shown for 12 quarters, although a short-term policy does not stretch that far. As the oscillatory modes have fairly long periodicities (cf. Section 5.7), rapid fluctuations are not present. There are, however, clear instances of 'overshoot' response.

There is a strong similarity between the diagrams for imports and for unemployment, variables that are related to the volume of economic activity. The instruments are thus less suited to reduce imports without causing unemployment. However, prices can be influenced separately from volumes, by using the indirect taxes instrument $\underline{TRK}$ in a sense opposite to that of $\underline{TRL}$ or $\underline{TRZ}$.

In the experiments, only the responses in the first three quarters are used. These are shown in Table 7.1 (p. 76).

The lags occurring in the responses are very evident. In the third quarter, when almost all instruments are showing some effect on almost all objectives, the parallelism between M and U, and between $\underline{TRL}$ and $\underline{TRZ}$ is apparent as well.

7.4. SPECIFICATION OF AN OBJECTIVE FUNCTION

7.4.1. *General*

The interrelations among economic variables, as hypothesized in the mathematical model in Chapter 4, make the policy problem complicated since a shift in one of the target variables by means of an instrument is, in general, accompanied by effects on other target variables. The plots of responses to the three tax instruments shown above indicate their divergence. Thus, if indirect taxes are manipulated, *ceteris paribus*, to shift prices, this will have effects on employment and balance-of-pay-

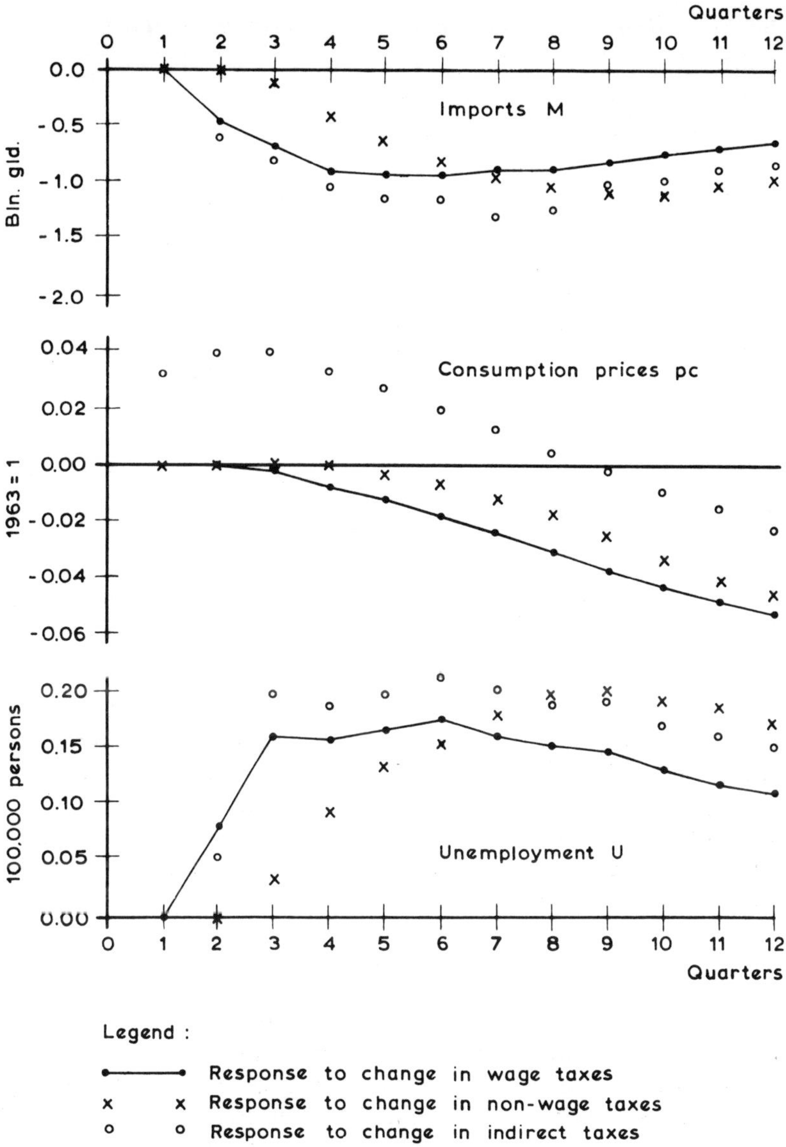

Figure 7.3. Response to unit step changes in instruments

Table 7.1. Response to unit step increases in instruments, 1963

Instruments	*Responses* M	pc	U
	First Quarter		
$\underline{TRL}$	0.000	0.000	0.000
$\underline{TRZ}$	0.000	0.000	0.000
$\underline{TRK}$	0.000	0.031	0.000
	Second Quarter		
$\underline{TRL}$	−0.458	0.000	0.080
$\underline{TRZ}$	0.000	0.000	0.000
$\underline{TRK}$	−0.572	0.039	0.100
	Third Quarter		
$\underline{TRL}$	−0.690	0.002	0.161
$\underline{TRZ}$	−0.171	0.000	0.030
$\underline{TRK}$	−0.821	0.039	0.195

ments. The difficulty in the achievement of objectives can be resolved by seeking an optimum that can be accomplished under the circumstances. To this end, the relative importance of different objectives has to be specified in quantitative terms. Similarly, the use of different instruments has to be limited because of the costs – administrative and psychological – involved in their use.

7.4.2. *Relative importance of policy objectives*

As described in Section 7.2, three desired time-paths – in balance-of-payments, price of consumption goods and unemployment – were chosen. To recollect, the balance-of-payments is regulated by regulation of imports because exports are exogenous in the macro-economic model.

On aprioristic grounds, the following events were regarded to be of equal importance:

1. A deviation from the desired value of the order of 0.5% of gross national product in imports M.
2. A deviation of 2% in prices pc from the desired $\tilde{pc}$.

3. A deviation from the desired level of 0.2 units (=20,000 persons) in unemployment U.

In the performance criterion J for the control problem in Section 2.2, a matrix V_i appears which weights the state vector values. The (3, 3)-submatrix of V_i corresponding to M, pc, and U, in this order, was selected to be

$$\begin{bmatrix} 1.0/(0.25\bar{Y})^2 & 0 & 0 \\ 0 & 1.00 & 0 \\ 0 & 0 & 0.01 \end{bmatrix}$$

The weighting matrix has one element, namely, (1, 1), which changes with time.

7.4.3. *Relative importance of the use of instruments*

The costs of using instruments are, likewise, given relative weights by considering the following events as equivalent to those given above:

4. A 5% autonomous change in the wage taxes from the reference level.
5. A 5% autonomous change in the non-wage taxes from the reference level.
6. A 10% autonomous change in the indirect taxes from the reference level.

The weighting matrix W_i, which gives relative importance to the use of instrument in the performance criterion J, can be specified as follows:

$$W_i = \begin{bmatrix} 1/(2.5\overline{TL})^2 & 0 & 0 \\ 0 & 1/(2.5\overline{TZ})^2 & 0 \\ 0 & 0 & 1/(5\overline{TK})^2 \end{bmatrix} \qquad (7.4.1)$$

The matrices W_i are specified once the reference values are given. They, however, change with the reference values.

7.4.4. *Choice of time-horizon and periods of instrument operation*

Consider a quarter serially indexed k. As explained in Chapter 5, the quarterly observations $y(1)$–$y(10)$ for this quarter become available in quarter $k+1$. The updating of the past estimates by means of the Kalman filter can then take place. The updating may indicate changes in instrument, or control action, values which can take effect in subsequent quarters $k+2$, $k+3$, ...

Because regulation in the short-run is of primary interest here, a nearby time-horizon suffices. Also, with transition of time, further control action is possible in the light of fresh information that becomes available in the mean while. From the response of the model to control actions, Figure 7.3, it is seen that, except for the influence of indirect taxes on price of consumption, effects in the quarter $k+2$ are absent if action is taken in that quarter. Therefore, to include measurable effects, the horizon is required to be sufficiently far away. On the other hand, a possible reversal of policy in quarter $k+3$ would affect quarter $k+5$, $k+6, \ldots$, and so on. As a compromise then the time-horizon was set at the end of quarter $k+4$, given that quarterly observations for quarter k are known in $k+1$.

Within this time-horizon, instruments can be given different values in periods $k+2$, $k+3$, and $k+4$. However, few, if any, governments would consider policies that changed every quarter. A minimum duration of two quarters for any policy appears practicable. Consequently, the instrument vector was set at equal values for quarters $k+2$ and $k+3$, and returned to zero in quarter $k+4$.

Based on these considerations, the objective function for quarter k can then be specified as

$$J = \min_{u_{k+1}, u_{k+2}} E\left[\sum_{i=k+2}^{k+4} \{(x_i - \tilde{x}_i)^T V_i (x_i - \tilde{x}_i) + u_{i-1}^T W_{i-1} u_{i-1}\}\right] \qquad (7.4.2)$$

where x_i denotes the desired state vector in quarter i and $u_{k+1} = u_{k+2}$ is a 3-dimensional vector having <u>TRL</u>, <u>TRZ</u>, and <u>TRK</u> as components. The non-zero elements of V_i are given above and they refer to elements (20, 20), (12, 12) and (8, 8) since imports, prices, and unemployment are the 20-th, 12-th and 8-th components of the state vector representing the economy. The matrices $W_{k+1} = W_{k+2}$ are those specified in Equation (7.4.1), and $W_{k+3} = 0$. It is important to keep in mind that the instruments are decided in quarter $k+1$ but take place in quarters $k+2$ and $k+3$; thus, a shift of one quarter is involved in the subscripts.

7.5. OPTIMAL FEEDBACK CONTROL ACTIONS

Making use of the results presented in Chapter 2, especially that the Certainty Equivalence Principle is applicable to the control problem, the control actions can be determined once the conditional expectation of the system state is known. In the expression for J given in Equation (7.4.2), the state vectors x_i can be replaced by their conditional expectations $\hat{x}_i$. In quarter $k+1$, when observations pertaining to quarter k are known, the updated estimate x_k^* can be determined. The predicted values of the system state based on information upto quarter k, replacing the random quantities by their expectations, are governed by the following set of equations.

$$\hat{x}_{k+1}=A_k x_k^* + B_k u_k \qquad (7.5.1)$$

$$\hat{x}_{k+2}=A_{k+1}\hat{x}_{k+1}+B_{k+1}u_{k+1} \qquad (7.5.2)$$

$$\hat{x}_{k+3}=A_{k+2}\hat{x}_{k+2}+B_{k+2}u_{k+2} \qquad (7.5.3)$$

$$\hat{x}_{k+4}=A_{k+3}\hat{x}_{k+3} \qquad (7.5.4)$$

As explained in the previous section, at time $k+1$ only $u_{k+1}=u_{k+2}$ are of interest; u_k cannot be changed and u_{k+3} is put at zero.

Let the notation $\hat{\hat{x}}_{k+2}$, $\hat{\hat{x}}_{k+3}$ and $\hat{\hat{x}}_{k+4}$ denote the predicted state vectors in the absence of u_{k+1} and u_{k+2}; that is,

$$\hat{x}_{k+1}=A_k x_k^* + B_k u_k \qquad (7.5.5)$$

$$\hat{\hat{x}}_{k+2}=A_{k+1}\hat{x}_{k+1} \qquad (7.5.6)$$

$$\hat{\hat{x}}_{k+3}=A_{k+2}\hat{\hat{x}}_{k+2} \qquad (7.5.7)$$

$$\hat{\hat{x}}_{k+4}=A_{k+3}\hat{\hat{x}}_{k+3} \qquad (7.5.8)$$

Equations (7.5.2)–(7.5.4) can then be expressed as

$$\hat{x}_{k+2}=\hat{\hat{x}}_{k+2}+B_{k+1}u_{k+1} \qquad (7.5.9)$$

$$\hat{x}_{k+3}=\hat{\hat{x}}_{k+3}+A_{k+2}B_{k+1}u_{k+1}+B_{k+2}u_{k+2} \qquad (7.5.10)$$

$$\hat{x}_{k+4}=\hat{\hat{x}}_{k+4}+A_{k+3}A_{k+2}B_{k+1}u_{k+1}+A_{k+3}B_{k+2}u_{k+2} \qquad (7.5.11)$$

The quadratic term involving $\hat{x}_{k+2}$ in the objective function J in Equation (7.4.2) can be written as

$$\begin{aligned}(\hat{x}_{k+2}-\tilde{x}_{k+2})^T V_{k+2}(\hat{x}_{k+2}-\tilde{x}_{k+2}) &= \\ =(\hat{\hat{x}}_{k+2}-\tilde{x}_{k+2})^T V_{k+2}(\hat{\hat{x}}_{k+2}-\tilde{x}_{k+2}) &+ \\ +2(\hat{\hat{x}}_{k+2}-\tilde{x}_{k+2})^T V_{k+2}(B_{k+1}u_{k+1}) &+ \\ +(B_{k+1}u_{k+1})^T V_{k+2}(B_{k+1}u_{k+1}) & \qquad (7.5.12)\end{aligned}$$

The first term on the right hand side in Equation (7.5.4) is a quadratic in deviations from the desired path and is independent of the instrument vector u_{k+1}. Therefore, it is not of interest so far as optimization with respect to the instruments is concerned. The second term is a product involving deviations from the desired path and the instruments. The last term is quadratic in the instruments. Therefore, the last two components of the sum are of interest.

In a similar fashion, the contribution of the other similar terms in J of Equation (7.4.2) can be expressed as sums of a quadratic term independent of instrument values, one involving products of deviations from the desired path and control values, and one quadratic in the control values. Finally, the objective function can be written in the form

$$J = K + 2u^T\beta + u^T M u \tag{7.5.13}$$

where K is a quadratic term independent of the control $u = u_{k+1} = u_{k+2}$, and the vector β and matrix M are given by the following relations:

$$\begin{aligned}\beta = {} & B_{k+1}^T V_{k+2}(\hat{\hat{x}}_{k+2} - \tilde{x}_{k+2}) + \\ & + (A_{k+2}B_{k+1} + B_{k+2})^T V_{k+3}(\hat{\hat{x}}_{k+3} - \tilde{x}_{k+3}) + \\ & + (A_{k+3}A_{k+2}B_{k+1} + A_{k+3}B_{k+2})^T V_{k+4}(\hat{\hat{x}}_{k+4} - \tilde{x}_{k+4})\end{aligned} \tag{7.5.14}$$

$$\begin{aligned}M = {} & B_{k+1}^T V_{k+2} B_{k+1} + W_{k+1} + (A_{k+1}B_{k+1} + B_{k+2})^T V_{k+3}(A_{k+1}B_{k+1} + \\ & + B_{k+2}) + W_{k+2} + \\ & + (A_{k+3}A_{k+2}B_{k+1} + A_{k+3}B_{k+2})^T V_{k+4}(A_{k+3}A_{k+2}B_{k+1} + \\ & + A_{k+3}B_{k+2})\end{aligned} \tag{7.5.15}$$

The optimal control action values are then found to be

$$u^* = -M^{-1}\beta \tag{7.5.16}$$

It can be remarked that only a weighted vector of deviations of the predicted values under assumption of no policy change, from the desired time-path, the vector β, enters into the determination of u^*.

The overall control strategy can now be summarized in the following steps:

Step 1. Updating the estimates x_k^* of the system state in the light of fresh observations.

Step 2. Prediction of the future values of the state vector under the assumption that no policy change takes place.

Step 3. Calculation of the weighted deviations of the predicted states, under assumption of no policy change, from the desired time-paths, Equation (7.5.14).

Step 4. Determination of the optimal control actions u^*, Equation (7.5.16).

Finally, the effect of the optimal control actions u^* on the value of the objective function can be evaluated. Substituting u^* in Equation (7.5.13), the optimal value of J is found to be

$$J^* = K - \beta^T M^{-1} \beta \tag{7.5.17}$$

In the absence of control action, the value of $J = K$. Thus, the effect of optimal control action is to diminish it by an amount $\beta^T u^*$.

7.6. NUMERICAL EXPERIMENTS

7.6.1. *General*

The procedure outlined in previous sections was used, in conjunction with the macro-economic model described in Chapter 4, to determine control actions which minimize the performance criterion J. The forty quarters of the period 1961–1970 were considered. The statistical data are described in Chapter 5. The exogenous variables were not forecast, as a factual application of the method would require. Instead, their constant smoothed values were employed. While this simplified computations, it also provided estimates closer to the later observations than conditional predictions would have been. However, in actual practice the prediction of exogenous variables would generally have profited from additional information and analysis. The derivation of the smoothed series from observations of the exogenous variables is indicated in Chapter 6.

The ten quarterly observation time-series $y(1)$–$y(10)$ are available in general within the next quarter. When they are known, conditional estimates and forecasts can be revised. Forecasts of the system state vector upto 4 quarters ahead are required for determining instrument values. Deviations from the desired path in the second, third and fourth quarters to come enter the objective function. The same control action is assumed to take place in the second and third quarter. It is computed by making use of Equations (7.5.14)–(7.5.16).

The numerical experiments are described in two stages. First, an experiment for a single quarter is described. Calculations for 36 quarters with a moving time-horizon of 4 quarters over the years 1961–1970 are then summarized. The effect of suppressing certain time-series out of $y(1)$–$y(10)$, representing non-availability of these observations, is examined in this case.

The updating of the conditional expectations was performed by means of the Kalman filter. The prediction-correction, or updating, matrix K in the Kalman filter algorithm had the fixed value, described in Section 5.6.

In each case, the initial state vector for the first quarter of 1961 was given the same value. This vector proved difficult to obtain. First, it was tried to let a repeated prediction and smoothing process converge numerically, in the same way as it was successfully performed for exogenous variables. It was found, however, that the matrix to be inverted in the Meditch smoothing algorithm was almost singular, so that very irregular results were obtained. Subsequently, it was attempted to minimize the sum of all 40 squared prediction errors in all the 10 observation series, which is a quadratic function of the initial state vector. The matrix to be inverted for this purpose again proved to be almost singular. This was redressed by fixing a small number of variables, including their related trends, and seasonals at zero. Suitable values for the remaining variables were then obtained.

7.6.2. *Experiment for a single quarter*

Consider quarter 2 of the year 1963. In this, the 10 quarterly observations of quarter 1 are available. The predictions of the value of imports (M), the consumption price (pc) and unemployment (U) in the quarters 3 and 4 of 1963 and quarter 1 of 1964 are given in Table 7.2. Corresponding desired values and deviations are also given in the table.

The value of the objective function J in the absence of control action was found to be 0.06476. The control action reduced it by an amount of 0.01628. Hence, the optimal objective function value, J^*, was 0.04848.

The optimal control action was found to be the vector [0.102, 0.097, 0.740]. Its three components are autonomous changes in wage, nonwage and indirect taxes; viz., $\underline{TRL}$, $\underline{TRZ}$, and $\underline{TRK}$, respectively. The units are bln. gld.

7.6.3. *Experiments for 36 quarters*

The optimization described above was repeatedly performed 36 times starting in 1961, quarter 1 and ending in quarter 4 of 1969. It thus involved a moving-horizon of 4 quarters in each case. The value of the objective function with and without the optimal control action and the effect of control action on it were computed in each quarter and added over the 36 quarters. The results are summarized in Table 7.3 (the meaning of the symbols $y(i)$ used is given in Section 5.3).

Table 7.2. Predictions and desired values based on quarterly observations upto quarter 1, 1963

		1963 Quarter 3	*1963 Quarter 4*	*1964 Quarter 1*
M	Predicted	18.642	18.739	22.411
	Desired	17.603	18.495	18.746
	Deviation	1.038	0.244	3.665
pc	Predicted	1.014	1.035	1.096
	Desired	1.037	1.050	1.055
	Deviation	−0.022	−0.014	0.040
U	Predicted	0.359	0.495	0.559
	Desired	0.446	0.448	0.452
	Deviation	−0.086	0.046	0.106

Table 7.3. Contributions of individual observation time-series

	Value of Objective Function in 36 Experiments		
	without Policy	*with Policy*	*Gain*
All Series	2.26	1.93	0.32
All Series Except:			
$y(1)$	4.90	4.39	0.51
$y(2)$	2.07	1.78	0.28
$y(3)$	2.55	2.18	0.37
$y(4)$	2.37	2.02	0.34
$y(5)$	2.99	2.52	0.47
$y(6)$	2.46	2.08	0.37
$y(7)$	28.61	13.11	15.49
$y(8)$	2.19	1.87	0.32
$y(9)$	1.99	1.71	0.27
$y(10)$	2.16	1.83	0.33

It is seen that $y(1)$, unemployment, and $y(7)$, the consumer price index, are more important than the other series. If either of these important series is left out, the predictions tend to diverge further from the desired values, and therefore a stronger policy is designed, with a corresponding greater gain, but still leaving a higher objective function value. It is perhaps not surprising that the series that are found to have the greatest effect are those that correspond directly to the target variables.

8. Conclusions

Macro-economic regulation is a complex control process. The insight about the functioning of the economy and the effect of the control actions is far from perfect, and economic phenomena are best considered as stochastic in nature. Observation of the economic activity and decision-making both involve some measure of approximation, arbitrariness and error. The execution of decisions could not be considered perfect either. Thus, the regulation of economic activity proceeds in uncertainty. In control theory, it can definitely be viewed as a stochastic control system.

Improvement of system performance when uncertainty and random disturbances are present is possible through closed-loop control policy. Observations of the system operation are used to determine appropriate control actions. In this study, a scheme is demonstrated, based on control theory concepts, in particular the closed-loop policy and the Kalman filter, which provides optimal regulation.

It employs a linear model, a quadratic objective function and its stochastic elements are additive. It can be shown that the Certainty Equivalence Principle is valid in this case. This result is more general than that encountered in econometric literature inasmuch as a closed-loop control policy is incorporated based on noisy observations of the system state, which may be fewer in number than the elements characterizing the state. Furthermore, it is shown that optimal control actions can be designed after predictions of the system state, based on latest information, have been obtained. These results provide a theoretical justification for the common practice of first obtaining predictions under the assumption of unchanged policy, and then designing regulatory measures as if the problem were deterministic.

The use of the Kalman filter enables one to deal appropriately with scarce and inexact statistical data. As it updates optimally the previous estimates on each occasion fresh information is available, the filter is convenient to use. Economy in computations is achieved because of its recursive nature. The contribution of all earlier information is summarized in the current estimates. The Kalman filter estimates are minimum-variance linear unbiased estimators of the system state, and hence have attractive properties.

The Kalman filter estimates enter into the determination of the optimal control actions for regulation. The filter weights the latest and the previous statistical data in a proportional-plus-integral scheme so that not only the latest but also the past information is appropriately incorporated. The resulting control actions are therefore smoother than in the case when only the latest information is taken into consideration. The proportional-plus-integral control policy thus adds to the general stability and continuity of system performance.

For the exogenous variables, a state space representation is provided generating time series with or without a seasonal component. This model satisfies controllability and observability requirements. Consequently, the associated filter is stable, and a finite steady-state error covariance matrix of the estimates is ensured. The corresponding smoothing matrix in the Meditch (1967) algorithm also acquires a steady-state value. The unknown initial vector containing trend and seasonal components can then be estimated by successive prediction, filtering, and smoothing over the sample of data. This treatment of time-series by the state-space methods of control theory may well have implications for time-series analysis in general.

The scheme described could well be used in the Netherlands. There exists a 'wiebeltax' specifically designed for economic regulation. Consequently, the instruments of economic policy can be quickly changed. A systematic quarterly analysis of the behaviour of the economic system already exists as well.

The numerical part of the study shows in a very simplified way how the scheme could have been applied in the Netherlands in the period 1961–1970. There appear to be no technical obstacles to doing the same thing on a larger scale and in a more realistic manner. It would, however, be helpful if a better econometric model were used, and the estimation of the initial vector requires more study.

References

Abert, J. G. (1969), *Economic Policy and Planning in the Netherlands, 1950–1965*, Yale University Press, New Haven.

Allen, R. G. D. (1967), *Macro-Economic Theory*, MacMillan and Co., London.

Aoki, M. (1967), *Optimization of Stochastic Systems*, Academic Press, New York.

Beld, C. A. van den (1968), 'An Experimental Medium-Term Macro-Model for the Dutch Economy' in Kendall, M. G. and others (Editors): *Mathematical Model Building in Economics and Industry*, Charles Griffin, London.

Bellman, R. (1957), *Dynamic Programming*, Princeton University Press, Princeton, New Jersey.

Blackman, R. B. and J. W. Tukey (1956), *The Measurement of Power Spectra*, Dover Publications, New York.

Bogaard, P. J. M. van den, and H. Theil (1959), 'Macrodynamic Policy-Making: An Application of Strategy and Certainty Equivalence Concepts to the Economy of the United States, 1933–1936', *Metroeconomica*, Vol. 11, pp. 149–167.

Boot, J. C. G., W. Feibes and J. H. C. Lisman (1967), 'Further Methods of Deviation of Quarterly Figures from Annual Data', *Applied Statistics*, Vol. 16, pp. 65–75.

Box, G. E. P. and G. M. Jenkins (1970), *Time Series Analysis: Forecasting and Control*, Holden-Day, San Francisco.

Brems, H. (1968), *Quantitative Economic Theory*, John Wiley, New York.

Brown, R. G. (1963), *Smoothing Forecasting and Prediction of Discrete Time Series*, Prentice-Hall, Englewood Cliffs, New Jersey.

Central Planning Bureau (1955), *Centraal Economisch Plan 1955*, Staatsuitgeverij, The Hague.

Central Planning Bureau (1961), *Central Economic Plan 1961*, Staatsuitgeverij, The Hague.

Deutsch, R. (1965), *Estimation Theory*, Prentice Hall, Englewood Cliffs, N.J.

Dhrymes, P. J. (1970), *Econometrics: Statistical Foundations and Applications*, Harper and Row, New York.

Driehuis, W. (1972), *Fluctuations and Growth in a Near Full Employment Economy*, Universitaire Pers Rotterdam.

Eijk, C. J. van and J. Sandee (1959), 'Quantitative Determination of an Optimum Economic Policy', *Econometrica*, Vol. 27, pp. 1–13.

Fel'dbaum, A. A. (1953), 'Optimal Processes in Automatic Control Systems', *Automat i Telemehan*, Vol. 14, pp. 712–728.

Fox, K. A., J. K. Sengupta and E. Thorbecke (1966), *The Theory of Quantitative Economic Policy with Applications to Economic Growth and Stabilization*, North-Holland Publ. Co., Amsterdam.

Freeman, H. (1965), *Discrete-Time Systems*, John Wiley, New York.

Goldberger, A. S. (1959), *Impact Multipliers and Dynamic Properties of the Klein-Goldberger Model*, North-Holland Publ. Co., Amsterdam.

Halanay, A. (1969), 'On the Controllability of Linear Difference Differential Systems', in H. W. Kuhn and G. P. Szegö (Editors): *Mathematical Systems Theory and Applications II*, Springer-Verlag, Berlin, pp. 329–336.

Holt, H. C., F. Modigliani, J. F. Muth and H. A. Simon (1960), *Production Planning, Inventories, and Work Force*, Prentice-Hall, Englewood Cliffs, N.J.

Kalman, R. E. (1960), 'A New Approach to Linear Filtering and Prediction Problems', Transactions of ASME, Series D, *Journal of Basic Engineering*, Vol. 82, pp. 35–45.

Kalman, R. E. and R. S. Bucy (1961), 'New Results in Linear Filtering and Prediction Theory', Transactions of ASME, Series D, *Journal of Basic Engineering*, Vol. 83, pp. 95–108.

Kalman, R. E. and R. W. Koepcke (1958), 'Optimal Synthesis of Linear Sampling Control Systems Using Generalized Performance Indexes', *Transactions ASME*, Vol. 80, pp. 1820–1826.

Kau, S. and K. S. P. Kumar (1969), 'Successive Linearization and Nonlinear Filtering', in M. Beckman and H. P. Künzi (Editors): *Computing Methods in Optimization Problems*, Springer-Verlag, Berlin.

Krasnakevich, J. R. and R. H. Haddad (1969), 'Generalized Prediction-Correction Estimation', *SIAM Journal on Control*, Vol. 7, pp. 469–511.

Luenberger, D. G. (1969), *Optimization by Vector Space Methods*, John Wiley, New York.

Meditch, J. S. (1967), 'Orthogonal Projection and Discrete Optimal Linear Smoothing', *SIAM Journal on Control*, Vol. 5, pp. 74–89.

Morgenstern, O. (1950), *On the Accuracy of Economic Observations*, Princeton University Press, Princeton.

Nerlove, M. and S. Wage (1964), 'On the Optimality of Adaptive Forecasting', *Management Science*, Vol. 10, pp. 207–224.

Phillips, A. W. (1954), 'Stabilization Policy in a Closed Economy', *Economic Journal*, Vol. 64, pp. 290–323.

Phillips, A. W. (1958), 'The Relation Between Unemployment and the Rate of Change of Money Wage Rates in the United Kingdom, 1861–1957', *Economica*, Vol. 25, pp. 283–299.

Raisbeek, G. (1964), *Information Theory: An Introduction for Scientists and Engineers*, MIT Press, Cambridge, Mass.

Shoeffler, S. (1955), *The Failures of Economics: A Diagnostic Study*, Harvard University Press, Cambridge, Mass.

Simon, H. A. (1956), 'Dynamic Programming Under Uncertainty with a Quadratic Criterion Function', *Econometrica*, Vol. 24, pp. 74–81.

Theil, H. (1964), *Optimal Decision Rules for Government and Industry*, North-Holland Publ. Co., Amsterdam.

Theil, H. and S. Wage (1964), 'Some Observations on Adaptive Forecasting', *Management Science*, Vol. 10, pp. 198–206.

Tinbergen, J. (1956), *Economic Policy: Principles and Design*, North-Holland Publ. Co., Amsterdam (Reprinted 1967).

Tustin, A. (1952), 'Feedback', in *Scientific American*, September 1952.

Vishwakarma, K. P. (1970a), 'Prediction of Economic Time-Series by Means of the Kalman Filter', *International Journal of Systems Sciences*, Vol. 1, pp. 25–32.

Vishwakarma, K. P. (1970b), 'Application of Kalman Filter Technique for Optimal Use of Time-Series Information in Dynamic Optimization Problems', Report 7018 of the Econometric Institute, Netherlands School of Economics, Rotterdam.

Wishart, D. M. G. (1969), 'A Survey of Control Theory', *Journal of Royal Statistical Society*, Series A, pp. 293–319.

Index